Compelling Form

Compelling Form:
Architecture as Visual Persuasion

By

J. Donald Ragsdale

CAMBRIDGE
SCHOLARS
PUBLISHING

Compelling Form: Architecture as Visual Persuasion,
by J. Donald Ragsdale

This book first published 2011

Cambridge Scholars Publishing

12 Back Chapman Street, Newcastle upon Tyne, NE6 2XX, UK

British Library Cataloguing in Publication Data
A catalogue record for this book is available from the British Library

ISBN (10): 1-4438-3286-3, ISBN (13): 978-1-4438-3286-1

TABLE OF CONTENTS

List of Figures vii

Acknowledgements xiii

Preface xv

Chapter One 1
Compelling Form: Architecture as Visual Persuasion

Chapter Two 18
Architecture: A Brief History of Elements

Chapter Three 38
Temples of the Muses

Chapter Four 62
Temples of the Performing Arts

Chapter Five 86
Houses of Worship: The Continental Christian Tradition

Chapter Six 113
Houses of Worship: The English-Speaking World

Chapter Seven 138
Universities as Visual Persuasion

Chapter Eight 166
Citadels of Power: National, State, Regional, and Local Government Buildings

Chapter Nine 195
The Domains of Royalty: Castles and Palaces

Chapter Ten 219
Places of Residence: Homes, Apartments, and Hotels

Chapter Eleven 242
The World of Commerce: Skyscrapers and Bridges

Chapter Twelve 265
Monuments and Memorials: The Synthesis of Art and Architecture
J. Donald Ragsdale and Frances E. Brandau-Brown

Afterword 288
Some Reflections on the Social Influence of Architecture

References 290

Index 294

List of Figures

2-1 Entablature, Pediment, and Doric Order, Bodleian Library 21
2-2 Ionic Order, Ashmolean Museum 23
2-3 Corinthian Order, The Circus, Bath, England 24
2-4 Hall of the Caryatids, Louvre 25
2-5 Ospedale degli Innocenti 32

3-1 Metropolitan Museum of Art 42
3-2 Philadelphia Museum of Art (under renovation) 43
3-3 Philadelphia Museum of Art, North Wing Tympanum 44
3-4 Kimbell Art Museum 45
3-5 Louvre Pyramid 46
3-6 British Museum 47
3-7 Tate Modern 50
3-8 Melbourne Museum 52
3-9 National Gallery of Victoria 53
3-10 Ian Potter Centre 54
3-11 Art Gallery of New South Wales 56
3-12 Museum of Contemporary Art, Sydney 57
3-13 Te Papa Tongarewa 58
3-14 Christchurch Art Gallery 60

4-1 Roman Amphitheatre, Lyon 63
4-2 Colosseum 63
4-3 Sydney Opera House 65
4-4 Gaertner Performing Arts Center 67
4-5 Gaertner Center Concert Hall 70
4-6 New Globe Theatre 71
4-7 Sheldonian Theatre 73
4-8 Royal Albert Hall 74
4-9 Banqueting House 75
4-10 Opéra Garnier 76
4-11 Opéra Garnier Interior 77
4-12 Victorian Arts Centre, Melbourne 78
4-13 Lincoln Center 79
4-14 Kennedy Center 81

4-15 Bass Concert Hall ... 82
4-16 Meyerson Symphony Center ... 83
4-17 Winspear Opera House ... 84

5-1 Notre-Dame de Paris ... 88
5-2 St. Denis ... 90
5-3 Notre-Dame de Chartres ... 91
5-4 Ste. Chapelle ... 92
5-5 Le Mont St. Michel ... 93
5-6 Notre-Dame de Reims ... 94
5-7 Cathedral of St. Michael and St. Gudula ... 95
5-8 Cathedral and Leaning Tower of Pisa ... 97
5-9 Duomo ... 98
5-10 Giotto's Campanile ... 99
5-11 Basilica of San Marco ... 100
5-12 St. Peter's Basilica ... 102
5-13 Santa Maria la Real de la Almudena ... 103
5-14 Cathedral of St. Vitus ... 104
5-15 Cathedral of St. Barbara ... 105
5-16 Esztergom Cathedral ... 106
5-17 Matthias Church ... 107
5-18 Vilnius Cathedral ... 108
5-19 Cathedral of St. Peter and St. Paul ... 109
5-20 St. Stephen's ... 110
5-21 Dresden Frauenkirche ... 111

6-1 Westminster Abbey ... 114
6-2 St. Paul's Cathedral ... 115
6-3 St. Martin in the Fields ... 116
6-4 Salisbury Cathedral ... 117
6-5 Salisbury Interior ... 118
6-6 Bath Abbey ... 119
6-7 Fan Vaulting, Bath Abbey ... 120
6-8 York Minster ... 121
6-9 Durham Cathedral ... 122
6-10 Cathedral of St. Giles ... 123
6-11 National Cathedral ... 125
6-12 St. Patrick's Cathedral ... 126
6-13 Mormon Temple ... 128
6-14 Grace Cathedral ... 129
6-15 St. Mary's Cathedral, Sydney ... 130

6-16 St. Andrew's Cathedral, Sydney ... 131
6-17 St. Patrick's Cathedral, Melbourne ... 132
6-18 St. Paul's Cathedral, Melbourne ... 133
6-19 Wellington Cathedral of St. Paul ... 134
6-20 Old St. Paul's Cathedral, Wellington ... 135
6-21 ChristChurch Cathedral ... 136

7-1 All Soul's College ... 142
7-2 Trinity College ... 143
7-3 Merton College ... 144
7-4 Bodleian Library ... 145
7-5 Radcliffe Camera ... 146
7-6 Ashmolean Museum ... 147
7-7 King's College Chapel ... 148
7-8 Senate House ... 149
7-9 Imperial College ... 150
7-10 Durham University ... 152
7-11 University of Edinburgh Old College ... 153
7-12 Sorbonne ... 154
7-13 Humboldt University ... 155
7-14 Harvard Yard ... 156
7-15 Massachusetts Hall ... 158
7-16 Carpenter Center for the Visual Arts ... 159
7-17 Texas A&M Approach ... 160
7-18 Austin Hall ... 162
7-19 RMIT ... 163
7-20 Rutherford House, Victoria University ... 164

8-1 Palace of Westminster ... 168
8-2 Scottish Parliament ... 170
8-3 Reichstag/Bundestag ... 172
8-4 Austrian Parliament ... 173
8-5 Hungarian Parliament ... 174
8-6 Congress of Deputies ... 175
8-7 US Capitol ... 176
8-8 US Supreme Court ... 177
8-9 Wellington Parliament and Beehive ... 178
8-10 Wellington Parliamentary Library ... 179
8-11 Parliament House of Victoria ... 180
8-12 State Library of Victoria ... 181
8-13 Ali 'iōlani Hale ... 182

8-14 Texas State Capitol 183
8-15 Bell County Courthouse 185
8-16 Hill County Courthouse 186
8-17 Ellis County Courthouse 187
8-18 Tarrant County Courthouse 188
8-19 Dallas County Courthouse 189
8-20 Denton County Courthouse 190
8-21 Brussels Town Hall 191
8-22 Vilnius Town Hall 192
8-23 London Town Hall 193

9-1 White Tower 196
9-2 Edinburgh Castle 197
9-3 Vincennes 199
9-4 Aigues-Mortes 200
9-5 Carcassonne 201
9-6 Hohensalzburg 201
9-7 Prague Castle 202
9-8 Castel sant'Angelo 203
9-9 Versailles 205
9-10 Hall of Mirrors 206
9-11 Fontainebleau 207
9-12 Belgian Royal Palace 208
9-13 Schönbrunn 209
9-14 Neuschwanstein 210
9-15 Hohenschwangau 211
9-16 Doge's Palace 212
9-17 Palacio Real Madrid 213
9-18 Buda Castle 214
9-19 Stockholm Royal Palace 215
9-20 Holyroodhouse Palace 216
9-21 Buckingham Palace 217

10-1 Mount Vernon 220
10-2 Monticello 222
10-3 The White House 224
10-4 Iolani Palace 225
10-5 Frick Mansion 226
10-6 Carlyle House 227
10-7 Lee Boyhood Home 228
10-8 Lee-Fendall House 229

10-9 Oak Alley ... 230
10-10 Shadows on the Teche ... 231
10-11 Oakwood Home ... 232
10-12 The Painted Ladies ... 233
10-13 The Circus ... 235
10-14 Royal Crescent ... 236
10-15 South Kensington Apartments ... 237
10-16 Place des Vosges ... 238
10-17 Grand Hotel Taipei ... 239
10-18 Grand Hotel Lobby ... 240
10-19 Adolphus Hotel ... 241

11-1 Flatiron Building ... 244
11-2 Chrysler Building ... 245
11-3 Empire State Building ... 246
11-4 Willis (Sears) Tower ... 247
11-5 Transamerica Building ... 248
11-6 Fountain Place ... 249
11-7 Bank of America Plaza ... 250
11-8 JP Morgan Chase Tower ... 251
11-9 D. R. Horton Tower ... 252
11-10 The Pickle ... 253
11-11 La Grande Arche ... 254
11-12 Pont du Gard ... 255
11-13 Bridge of Sighs ... 256
11-14 Rialto Bridge ... 257
11-15 Tower Bridge ... 258
11-16 Charles Bridge ... 258
11-17 Chain Bridge ... 259
11-18 Brooklyn Bridge ... 260
11-19 Golden Gate Bridge ... 261
11-20 Sydney Harbor Bridge ... 261
11-21 Victoria Station ... 262
11-22 Flinders Street Station ... 263

12-1 Lincoln Memorial ... 267
12-2 Jefferson Memorial ... 268
12-3 Albert Memorial ... 269
12-4 Queen Victoria Monument ... 270
12-5 Scott Memorial ... 270
12-6 Vittorio Emanuele II Monument ... 271

12-7 Chiang Kai-Shek Memorial ... 272
12-8 Washington Monument ... 273
12-9 Vietnam Veterans Memorial ... 274
12-10 World War II Memorial ... 275
12-11 Arizona Memorial ... 276
12-12 Eiffel Tower ... 277
12-13 Arc de Triomphe ... 279
12-14 Memorial to the Murdered Jews of Europe ... 280
12-15 Korean War Veterans Memorial ... 281
12-16 FDR Memorial ... 282
12-17 Trajan's Column ... 283
12-18 Siegessäule ... 284
12-19 Brandenburg Gate ... 285
12-20 Nelson Column ... 286
12-21 Statue of Liberty ... 287

Some figures were obtained from commons.wikimedia.org and were contributed to this public domain depository by the following:

4-13 Lincoln Center, Matthew Bisanz, 18 July 2010.
4-14 Kennedy Center, Steve, 20 May 2007.
5-4 St. Chapelle, Beckstett, 2005.
5-5 Le Mont St. Michel, Eric Pouhier, 2005.
5-11 Basilica of San Marco, Marcus Mark, 2006.
6-11 National Cathedral, Mariordo, 24 August 2009.
8-7 US Capitol, Kevin McCoy, 4 May 2004.
8-8 US Supreme Court, D. B. King, 19 September 2004.
9-11 Fontainebleau, Carolus, 2 August 2006.
10-1 Mount Vernon, Ad Meskens, July 2007.
10-2 Monticello, Christopher Hollis for Wdwic Pictures, 1 July 2001.
10-9 Oak Alley, Jeffreyareed, 2009.
10-10 Shadows on the Teche, Carol M. Highsmith, 16 March 2007.
11-18 Brooklyn Bridge, Tiago Fioreze, 6 June 2009.
12-9 Vietnam Veterans Memorial, Lorax, 2003.
12-11 Arizona Memorial, Stan Shebs, 2003.

Figures 9-3, Vincennes, 9-9, Versailles, and 9-10, Hall of Mirrors are by Sandy Irwin Ragsdale. All other images are by the author.

ACKNOWLEDGEMENTS

As in my previous three volumes on visual persuasion published by Cambridge Scholars Publishing, I wish to acknowledge my debt to a number of people. There are the historical influences: my college and graduate school professors, my chairs and deans in various places of employment, and my professional colleagues. Then there are the vital current ones: John M. de Castro, Dean of the College of Humanities and Social Sciences at Sam Houston State University, who has fully supported my work both morally and financially; my colleagues in the Department of Communication Studies—Rick Bello, Frances Brandau-Brown, Terry Thibodeaux, and Shaun Zhang—with whom I enjoy a very supportive environment and with whom I have collaborated on this and other research projects; Andy Nercessian and Amanda Millar at Cambridge Scholars Publishing who have respectively endorsed my ideas and edited my manuscripts; and my family—Sandy, Paul, Alan, Katie, Mike, Amy, Sydney, and Tristen. Thank you one and all.

A version of the first chapter of this book has appeared previously: Ragsdale, J. D., 2010. Compelling form: Architecture as visual persuasion, *The International Journal of the Arts in Society,* 5,89-102, published by Common Ground Publishing. Chapter Twelve, as noted in the book itself, was co-authored with my colleague Frances E. Brandau-Brown.

PREFACE

My earliest career plan as a child was to be an architect. While that intention did not ultimately come to fruition, I recall much time spent as a boy drawing floor plans and planning my dream home. I have retained over the years an abiding interest in buildings and structures in general. Although I chose the field of communication studies and a career as a professor, it turns out that I have now found a way to merge my original and my eventual interests through the study of visual persuasion. This book is my fourth published by Cambridge Scholars Publishing on visual persuasion. Each has considered structures as types of visual influence. The first was a general and wide-ranging study of structures as argument (Ragsdale 2007). The second and third were studies of American and Western European museums. This one is a comprehensive study of many forms of architecture as examples of the compelling power of structures.

This consideration of the compelling forms of architecture was a daunting one in terms of its sheer scope. Even so, this book has very clear limitations. First, this is a book about Western architecture. In spite of its importance, architecture in the Nonwestern tradition was beyond the scope of the book, although I am confident that the principles on which I base my findings in this book are equally applicable to the Nonwestern tradition. I am simply limited in how much of that tradition I have experienced and studied. Second, and coincident with the first, I have restricted my assessments, with very few exceptions, to that architecture which I have actually visited, studied, and photographed. With widely available written and electronic sources, it would be possible to write about architecture without having seen it, but in my judgment it would not be desirable to do so. If one is to assess a building as a type of visual persuasion, then it seems necessarily that one should be able to draw upon personal experience as well as that of others.

The reader will notice, third, that there are both curious omissions and curious inclusions among the architectural items. Both of these facts follow from my desire to study forms that I have experienced. Some of my research for this book was done over many years in a casual way, without the thought that it would be used in a book. More recently, I have made several deliberate research trips to study and photograph architecture, but some desirable inclusions eluded me for the sake of time and money.

Finally, the reader will notice that some of the figures included are not of the highest photographic quality. This is the result of my own, perhaps perverse, desire to use my own work if at all possible. When that work did not meet minimal standards for presentation, I resorted to the use of photographs in the public domain, and these are identified at the end of the list of figures. Three photographs are by my spouse, Sandy, and are also identified in this list.

This work is meant as a scholarly one. I consider my task to have been an effort to demonstrate that the elements of visual literacy and the use of these elements in patterns of design were intended to strike the viewer as compelling. I also think that this impact is not unlike that which viewers call beautiful, and that thought is no different than Vitruvius's (1960) idea that architecture has three components: strength, utility, and beauty. However, the book need not be thought of as scholarship only, nor should it be avoided by the layperson. It is not really a travel guide, but it surely should serve well the visitor to any of the museums, performance halls, cathedrals, castles, palaces, and the like described herein.

CHAPTER ONE

COMPELLING FORM: ARCHITECTURE AS VISUAL PERSUASION

The traditional view is that communication is predominantly a human activity, and the emphasis has usually been on speakers and their messages. To be sure, scholars have understood that such other things as audiences and occasions made a difference in the outcome of communication, but in both rhetorical criticism and teaching it has been the speaker and her or his message that were spotlighted. Today it is a commonplace that communication is a process rather than a set of static and disconnected elements or something that requires looking only at one or two of the elements. One cannot understand the process without accounting for all of the elements, their interconnections, and the influence of the context.

One of the simplest representations of the process of communication, popularly referred to as the SMCR model (Berlo 1960), emphasizes the interdependence of a source or sources (S), a message or messages (M), a channel or channels (C) of communication, such as the vocal-auditory channel of spoken discourse, and a receiver or receivers (R) of the message or messages. Each of these elements influences and is influenced by the others, and of course communication always occurs within a particular context or in a particular situation.

The literary critic Kenneth Burke (1945) proposed a similar way of looking at human communication, which has gained wide currency particularly in the sub-area of communication studies called rhetorical theory. Burke used an analogy from drama in proposing what he called a dramatistic pentad, which would account for the process of communication. It included five elements: act, agent, agency, scene, and purpose. While there is nothing in the SMCR model that confines it to the typical act of communication, a speech, the Burkean model has been more successful in being extended to different types of communication. An act may be any purposive behavior and may not be confined to human beings. An agency would normally be a human responsible for the act but would not have to be. The agency is the means by which an act is performed. As we shall see, agency may not be done quickly but might be developed over a great span of time by different, unrelated agents. The scene is the context or environment of the act, and the purpose recognizes the importance of intent in the creation of acts or messages.

The idea of a typical act of communication today is not especially helpful in understanding the diversity of forms used by contemporary media and communicators. With the advent of visual communication as a legitimate area of study, it is almost the case that atypical has become typical. There is, for example, little question anymore that structures, buildings of all kinds, are communicative even persuasive (Ragsdale, 2007, 2009a, 2009b). What this means at its heart is that architecture is not some monolithic riddle, nor is it a meaningless accumulation of structural material. Architecture *means* something. While it might be superficially tempting to think of a building as analogous to a speaker in the process of communication, it is rather the architect that is the speaker's analogue. Architecture itself is a message or an agency. As John Ruskin (1989, 8) said, "All architecture proposes an effect on the human mind, not merely a service to the human frame."

Saying that architecture *means* something places it within the purview of semiotics or semantics. Semiotics or semantics is the study of meaning—how something means, the nature of meaning, the importance of types of meaning, and the like. In previous accounts of structures and museums as visual persuasion (Ragsdale 2007, 2009a, 2009b), the semiotic system of Charles S. Pierce (Buchler 1955) as adapted by Paul Messaris (1997) was used to provide a typology of visual signs. In this system, there are two types of visual signs: icons and indexes. Icons are *representational,* in that they look like that for which they stand. Icons are abundant in the portal sculptures of Gothic cathedrals. Indexes are *documentary*. They are direct evidence of a thing, such as unaltered photographs and artifacts like a tank used in the occupation of Budapest. Messaris recognized also that how a sign is interpreted often depended upon its juxtaposition to other signs. The theory of *montage* in film is an example of this juxtapositional effect. Messaris termed this phenomenon syntactic indeterminacy, and it has been used to explain such persuasive effects as those of the great dinosaur skeleton in the rotunda of The British Museum of Natural History (Ragsdale 2007) and the location in the Louvre of *Nike of Samothrace* at the head of a grand staircase (Ragsdale 2009b).

It was necessary to expand this simple system to account for museums and museum contents (Ragsdale 2007). The basic system did not, for example, afford a way to assess the difference between *discursive* and *nondiscursive* visual signs. Discursive signs have a clear parallel in language, as in the case of the "sermons in stone" of the portal iconography at Notre-Dame de Paris. However, the light-gathering structure of a cathedral, probably intended to represent the nature of God himself or of Heaven, seemed not to have any parallel in language. The term nondiscursive was used to describe such visual elements. Color is a nondiscursive sign. Space is similarly a nondiscursive sign. The ruins at Oradour-sur-Glane near Limoges in France and the tranquility of a Shinto shrine must be considered within their encompassing contexts. It was also necessary to resort to the distinction between central and peripheral pathways made in the Elaboration Likelihood Model of how persuasive messages are processed (Petty and Cacioppo 1986). Central processing involves deliberative thought about the content of a message, while peripheral processing is more impressionistic. This distinction was especially useful in describing the differences between discursive and nondiscursive visual signs. Discursive ones would seem to necessitate central processing, while nondiscursive ones are likely to be processed peripherally.

It was also necessary to add to icons and indexes another pair of visual signs to account for the varieties of art found in museums (Ragsdale 2009a). There are inadequacies in the icon/index categorization raised by the existence in the vast majority of art objects of symbolism. A recognition of such a symbolic image is to be found in C. S. Peirce's theory of verbal signs. For Peirce, a symbol was a conventionalized representation of its object. The best example is a word, such as "man." By contrast, a stick figure drawing of a man is an icon, and a photograph of a man is an index. What, then, is a visual symbol in Peircean terms? While it might be more difficult in visual communication to find instances than in verbal communication, where symbols abound, there are conventionalized visual symbols to be found in every culture (Ragsdale 2007). Trees of life, owls, crucifixes, angels, and the like all have implications beyond mere representation. Accepting that visual signs may be symbolic provides a means for interpreting the many instances of traditional art, which are merely puzzling when contemplated as icons.

Peirce did not anticipate the rise of modern art and its practical rejection of representation. The works of such painters as Pollock, Kandinsky, and Mondrian cannot be classified as either icons or indexes. Neither can they be called symbols. For this reason, a fourth sign is required, which has been called a *presentation*. A presentation is a visual composition of dots, lines, tones, colors, and the like presented through visual communicative strategies such as balance, symmetry, sharpness, randomness, and the like, but it does not "represent" anything. The interpretation of a presentation will always be subject to more variation than that of a symbol, an icon, or an index, since presentations do not have clearly differentiated objects. Objects of presentations are ordinarily unknowable, ineffable, or indecipherable, and their persuasive impact is likely to be a function of subjective interpretation.

What is architecture, then, as one of the types of visual signs just mentioned? Is it an icon? Is it an index? Is it a symbol? Is it a presentation? Is it a combination of signs? Clearly, buildings are not indexes, although other structures such as bridges probably are, neither do they match the definition of an icon. Aspects of some buildings are certainly symbolic, as in the case of the cruciform floor plan of a Gothic cathedral, but taken as a whole it does not seem reasonable to call buildings symbols. While buildings certainly contain a variety of visual elements and communication strategies, as will subsequently be evident, it does not seem productive to think of them as combinations of signs. How, then, does a building mean? The most appealing answer seems to be that buildings are presentations. They depend for their meaning on the visual elements and communication techniques which are their basic components. While parts of some buildings may be discursive, architecture is probably best conceived as a nondiscursive visual sign and likely one that is processed peripherally.

Like the other arts, architecture has the power for its message or messages to be repeated again and again across the ages. The other arts, however, do not have quite the omnipresence of architecture. It is in this sense that Roth (1993, 1) calls architecture "the unavoidable art." "Every moment, awake or asleep," Roth writes, "we are in buildings, around buildings, in spaces defined by buildings, or in landscapes shaped by human artifice. It is possible to take deliberate steps to avoid looking at painting, sculpture, drawings, or any of the other visual arts, but architecture constantly touches us, shapes our behavior, and conditions our psychological mood." "Architecture is shelter," Roth continues (141), "but it is also symbol and a form of communication. . . . [and] is a physical representation of human thought and aspiration, a record of the beliefs and values of the culture that produces it."

It is the omnipresence and enduring nature of architecture and their potential influence that makes it very important indeed to understand how communication works in this case. This is especially so to the degree that buildings also persuade or exert social influence. *Structures as Argument* (Ragsdale 2007) was an initial exploration of buildings as means of persuasion. It examined museums, zoos, cathedrals, Spanish missions, Protestant churches, non-Western buildings, monuments, and cemetery stones, and found clear evidence of the use of architecture to exert influence. *American Museums and the Persuasive Impulse* (Ragsdale 2009a) and *Western European Museums and Visual Persuasion* (Ragsdale 2009b) extended this study of buildings to an assessment of museums, their collections, their exhibition designs, and their architecture, as visually persuasive phenomena.

The present volume is an extension of the previous three and grows directly out of the approach used before. It develops a methodology for assessing architecture as communication and as persuasion, beginning with the fundamental elements and communication principles of visual literacy. It then considers the specific elements of architecture in terms of these basic ideas to construct a framework for evaluating the visual experience of buildings. This framework, which is not unlike a set of distinctive features, then will be used to assess the messages that are cathedrals, museums, performance halls, government buildings, skyscrapers, and the like, as kinds of visual communication and persuasion. Before proceeding further, it would be wise to consider the differences between communication and persuasion as those terms are used here.

Communication and persuasion, while not synonyms, are considerably more alike than they are different. Since the process of communication was introduced at the outset of this chapter, it is really only necessary to consider persuasion. Communication is the umbrella term, and persuasion is one of its forms. Indeed, persuasion is probably the most important of the forms of communication. It refers to messages that are designed to change attitudes, beliefs, values, and behavior, although it is certainly true that even messages not so designed may have a persuasive effect. Mere awareness of the availability of a product may lead to a decision to purchase it even in the absence of an advertising campaign. It should be evident in the forthcoming pages that buildings are normally persuasive, not just communicative. Roth (1993, 4) puts it this way: "architecture has the power to affect and condition human behavior; the color of walls in a room, for example, can help to determine our mood. Architecture acts upon us." What this chapter is primarily about is finding ways to understand how architecture acts upon us and to assess the effectiveness of its social influence.

It has long been the traditional view that persuasion results from the use of modes of proof—logic, emotional appeals, and the source's credibility. Logic or reason provided the basis for insisting that an attitude or an action needed to be changed, with the premises of such reasoning amply supported by examples, statistics, facts, testimony, and the like. Appeals to patriotism, to religious beliefs, or even to fear provided the motivation for action. Sometimes, perhaps even usually, the source's reputation for expertise, trustworthiness, and good will added strength to the persuasive effort. Sometimes, one mode of proof alone might be persuasive. At other times, two or three modes might be necessary. What is obvious, however, is that there is nothing directly comparable to this traditional view of persuasion to be found in the messages that are buildings. If it is not modes of proof that influence our attitudes and behavior when we encounter buildings, then what is it that does? The answer to this question lies in an analysis of the elements of architecture, which are in fact what we see. An understanding of these elements in turn begins with the most basic elements of visual communication. These ideas have been explored at length in previous works (Ragsdale 2009a, 2009b), and the following discussion repeats those earlier explorations.

The Concept of Visual Literacy

Visual literacy is a trained awareness of the elements of visual communication. A student of architecture must possess this awareness. Theories of architecture, however, begin at a macro rather than a micro level. In the principal document from antiquity on architecture available today, *Ten Books on Architecture* (1960), Vitruvius describes the subject as having three guiding principles: strength (*firmitas*), utility (*utilitas*), and beauty (*venustas*). In Chapter 2 of Book 1, he sets forth "The Fundamental Principles of Architecture" (13). These are order, arrangement, eurythmy, symmetry, propriety, and economy. As with many ancient texts, these terms are not so clear today as they undoubtedly were in Rome. Order, for example, appears to arise when the parts or modules of a building are all in harmony with each other and when they are symmetrical and in proper proportion to one another. The idea of thinking of a building as a set of modules rather than as an organic whole seems quaint today, but it is not an irrelevant way of thinking.

Arrangement proceeds directly from order and has to do with "the putting of things in their proper places" (Vitruvius 1960, 13). "Eurythmy is beauty and fitness in the adjustments of the members" (14). This beauty and fitness come about when a building's dimensions are properly proportioned. Eurythmy also requires symmetry, which Vitruvius includes as a separate principle. Appropriateness was an important consideration in many of the Classical arts, such as oratory, and architecture is no exception. The Greeks, and later the Romans, thought of art as arising from adherence to a set of rules. The Greek word was *techne.* Vitruvius explains that propriety comes from "prescription . . . usage . . . or from nature." One particularly helpful example he gives is that a building with an elegant interior also ought to have an elegant entrance. Economy, finally, has to do with costs and the use of materials.

To be sure, Vitruvius's ten books are quite explicit in explaining these basic principles, but the list itself remains general. Modern authors fortunately provide more elaborated lists. Strickland (2001, xi-xiv), for example, includes the following:

- Rhythm
- Line
- Scale or size
- Light
- Texture
- Color
- Ornament
- Acoustics
- Site
- Space
- Weight and Mass

Roth (1993) provides a very similar list:

- Proportion
- Scale
- Rhythm
- Texture
- Light
- Color
- Ugliness
- Ornament
- Acoustics
- Space
- Function
- "Firmeness"

These two lists are not substantively different, which will subsequently make it possible to combine them into a framework for assessing the visual impact of architecture.

Underlying the three principles of Vitruvius and the lists of Strickland and Roth are some even more basic elements of visual literacy, a micro level as it were. As sounds or phones underlie spoken language, so "the dot, line, shape, direction, tone, color, texture, dimension, scale, movement" underlie visual communication (Dondis 1973, 39). These basic elements are used to formulate visual messages through the application of "visual techniques" or "communication strategies" (104). Dondis (1973, 16) discusses a number of the most common of these communication strategies but points out that there are undoubtedly many more. Some of the most common are as follows:

- Balance—Instability
- Symmetry—Asymmetry
- Regularity—Irregularity
- Simplicity—Complexity
- Unity—Fragmentation
- Economy—Intricacy
- Understatement—Exaggeration
- Predictability—Spontaneity
- Boldness—Subtlety
- Simplicity—Complexity
- Depth—Flatness
- Sharpness—Diffusion
- Sequentiality—Randomness
- Repetition—Episodicity

Visual literacy, then, requires an awareness of the fundamental means of visual communication, i.e., the component dots, lines, colors, and so on, but it also requires an appreciation of the ways in which these components may be put together to create such effects as depth, subtlety, sharpness, balance, and the like.

Ching (1973) is well aware of the interaction of these two levels in his account of the art of architecture. He begins his account with the most basic elements: points, lines, planes, and volume. Ching then shows how form and space is organized according to the principles of circulation, proportion, and scale acknowledging a familiar set of principles: an axis, symmetry, hierarchy, rhythm/repetition, a datum, and transformation. While his list of principles is shorter than those of Strickland and Roth, Ching shows more explicitly than either of the other two writers how such elements of architecture as rhythm, line, and scale are related to the more fundamental basic elements and communication strategies.

Another and perhaps more enlightening way of regarding this third level of elements is to see it as describing the goals or outcomes of architecture. The architect, for example, manipulates such things as lines, shapes, tones, and textures using principles of balance, symmetry, and the like to arrive at a structure that is firm and strong, functional, and comely to look at. Turned back on themselves, the elements of architecture offer a framework for the assessment of structures as forms of visual communication and persuasion. In order to build a clear framework for this assessment, it is necessary to examine each of the elements of architecture in terms of their basic visual components. Essentially, one wants to know how meaning is conveyed by architecture and how that in turn results in social influence.

Communication through Architecture

Perhaps the best-known and most influential treatise on architecture ever written is Vitruvius's *Ten Books on Architecture* (1960). Vitruvius is spare indeed in enumerating the basic elements of his subject, saying that architecture should be strong, functional, and beautiful (1914, 17). He considers strength at considerable length and function a bit less so. While beauty might seem to students of visual communication to be the most significant of the three elements, Vitruvius is virtually silent on its components. However, beauty is rather obviously a persuasive goal, inasmuch as to regard an object as beautiful is to have been compellingly influenced by its form (Ragsdale 2008a). Indeed, it is implied in the lists of elements of architecture of Strickland and Roth mentioned above that the ultimate goal is primarily to influence the observer. To begin, then, one must scrutinize each element first as a means of communication.

As Roth (1993, 57) states outright, these elements arise from "the most fundamental concept . . . that the mind, particularly the human mind, is programmed to seek meaning and significance in all sensory information sent to it. . . . [which] no doubt, is linked with the instinct for survival." Additionally, "how the mind interprets forms and patterns presented to it is the subject of Gestalt psychology (from the German *Gestalt,* 'form' or 'shape')." The mind does this interpretation through a set of preferences "for proximity, repetition, simplest and largest figure, continuity and closure, and figure/ground relationship" (59). Roth's explanation is not unique, since Dondis (1973) also bases her discussion of the elements and communication strategies of visual literacy on the perceptual principles of Gestalt psychology. In viewing works of art, for example, Dondis asserts that one's eye is naturally drawn to "the felt axis . . . in an unending process of establishing the relative balance" (27). Next, the eye is drawn to the lower half of the visual field. Elements of architecture, it would seem to follow, are consciously selected principles whose use produces a predictable effect. Let us see how this idea plays out in practice.

Strickland's (2001) first element is rhythm, which is also included in Roth's (1993) list. Rhythm is "a pattern of . . . repetition of elements like solids and voids, walls and windows, projecting or receding parts" (Strickland 2001, xi). It is "the alternation between incident and interval, between solids and voids" (Roth 1993, 67). "This architectural rhythm is read by scanning the surface, much as one might scan a musical score. . . .

for both must be experienced in time." This account suggests flat surfaces, but Roth says that "undulating or curving walls" also qualify, with the rhythm expressed as "curves and countercurves" (70). In terms of the communication strategies of visual literacy, rhythm comes from the use of sequentiality, regularity, and, obviously, repetition. The experience of the rhythmic patterns of a building is a pleasing one based as it is on repetition and continuity.

Rhythm is such a fundamental characteristic of architecture that there are innumerable examples to be found. The entry façade of the Prado in Madrid is a particularly nice example. The entry itself is a projecting portico supported by a row of six Doric columns. The ground floor is a series of arches interspersed with solid supporting walls within which are niches for sculpture. The second floor is a succession of Ionic columns and windows. The use of successive columns is a very old and frequently repeated feature, perhaps best illustrated by the Parthenon in Athens. The Altes Museum on Berlin's Museumsinsel features Ionic columns each topped with the German eagle. An expanse of columns also is evidence of an underlying solid structure of considerable strength.

Roth's (1993) first element is proportion. Although Strickland (2001) does not include this term in her list, it is clearly of fundamental importance in the design of buildings. The primary reason for this is that something which is out of proportion will strike the eye as irregular or out of balance. "The mind . . . seeks out mathematical and geometrical relationships—or proportions—in patterns" (Roth 1993, 62). Architecture in Classical Greece often utilized the Golden Section, a "proportional relationship . . . believed to be perfect." Such a proportion obtains in a rectangle, for example, when the shorter side of the shape is to the longer side as the latter is to the longer and the shorter sides combined.

The Greeks regarded the human body as a divine proportion created in the image of the gods. Vitruvius (1960) explained this proportion by setting the center of the geometric pattern as the navel and then showing that the outstretched limbs described both a square and a circle, "the most basic and ideal of geometric figures" (Roth 1993, 62). Leonardo da Vinci made a drawing, much copied even today, of "Vitruvian man," showing these relationships. While modern anatomists would warn against any idea of a uniform proportion among human beings, it is very curious indeed that several writers on the subject of facial attractiveness seem to agree that a vertically symmetrical face is universally regarded as more attractive than an asymmetrical one (Rhodes, Proffitt, Grady, and Sumich 1998).

In buildings as well, symmetry is of paramount importance. It is a communication technique in Dondis's (1973) system and underlies the rhythmic use of columns, arches, and statues in the Prado and columns in the Altes Museum. The Pantheon in Rome is a building of enduring beauty, probably especially because it uses both the square and the circle combined together in its floor plan. The design is also fundamentally strong. Closely related to the idea of proportion is line (Strickland 2001). The line or lines of a building suggest a number of visual qualities. Horizontal lines imply solidity, while vertical ones imply "uplift" and "striving for ascent to a heavenly realm" (xii). Diagonal lines are dynamic and "imply action and energy." Lines, of course, are among the basic elements of visual literacy and are fundamental to the formation of geometric shapes.

Scale or size is the next element of architectural design to consider, and it appears in both Strickland's and Roth's list. In thinking of the great works of architecture, it is tempting to think that the larger a building the better. However, it is just as true that a smaller structure that is appropriate to its site and is suitable for its function deserves equal plaudits. In general, however, the structures to be considered in this book are large, even monumental, ones. Considered in relation to one's own size, large scale structures can be overwhelming or intimidating. Rome's Colosseum is such a place. However, the sheer scale of some buildings, like the scale of the Grand Canyon of the Colorado, also renders them awe-inspiring and visually arresting. The chateau and grounds at Versailles come to mind as does Schönbrunn Palace in Vienna. St. Peter's Basilica in the Vatican, La Sagrada Familia in Barcelona, and Salisbury Cathedral in England are examples, as is the American Capitol Building in Washington, DC, the skyscrapers of Manhattan, and La Défense in Paris.

Following scale or size, light is of crucial importance in the impact a building may have on the viewer, especially in the interior. Light enhances the perception of openness, spaciousness, and freedom of movement. The great Gothic cathedrals differed from their Romanesque predecessors in no greater way than their light gathering construction. Light has a fundamental symbolic value in the Christian faith, with darkness associated with evil and wrongdoing. Museums use skylights wherever possible to facilitate the viewing of their collections. Of course, there is also artificial light, which architects carefully utilize whenever natural lighting is not available. Light interacts with two other architectural elements: space and color.

The experience of space or spaciousness is directly influenced by lighting levels, so the architect must be aware of how the two interact as a design is being formulated. Roth (1993) devotes an entire chapter to space. He identifies four kinds of space with respect to a structure's interior. The first is the obvious physical space, that which is enclosed within a building. The second is perceptual space or "the space that can be perceived or seen" (45). Third there is conceptual space, "which can be defined as the mental map we carry around in our heads, the plan stored in our memory. Buildings that work well are those that users can grasp easily in their mind's eye and in which they can move about easily." Fourth, there is behavioral space, "or the space we can actually move through and use." Although focused on buildings themselves, Roth points out that these four types of space could easily be adapted to such outdoor areas as a town square or the courtyard of a cathedral. As one thinks about how to assess the visual effects of architecture, this enlarged description of space should offer helpful guidelines.

With respect to behavioral space, architects may design buildings so that movement between rooms is either facilitated or not, which Roth (1993) calls interwoven versus static spaces. In connection with such movement, "space can determine or suggest patterns of behavior by its very configuration, regardless of barriers or hindrances. We speak of **directional space**, as distinct from **nondirectional space**" (51). Some designs suggest paths through space, such as the design of a contemporary shopping mall does to influence which shops will draw the most customers. Buildings, of course, cannot properly be thought of divorced from their surroundings. Sometimes those surroundings are as important in the visual experience as the structure itself. Site, therefore, is an independent element of architectural design.

In previous studies of the visual persuasiveness of museums, site was a vital aspect of assessing the power of a building's influence (Ragsdale 2009a, 2009b). Traditionally, art museums were thought of as temples of the muses. As a result, the ideal location for an art museum was on a hill, so that the ascent mimicked the approach to the realm of the gods. One, of course, recalls the Biblical reference to "a city set upon a hill," and many buildings are located so as to give them majesty. The Parthenon, set on the hill of the Acropolis in Athens, may be the best known example. When a hill is not available or appropriate, then the architect may use a wide avenue or the like. The Arc de Triomphe is located at one end of the Champs Élysées, with the Louvre at the other end, and west along the same axis is La Défense. These locations are hardly coincidental, and one can easily find numerous examples of the placement of museums, monuments, government buildings, cathedrals, and so on. Site enhances the visual impact of a structure and also draws visual attention through the natural tendency of the human eye to follow horizontal lines.

Light also augments the element of color in visual design. Color itself has specific visual impacts which are typically quite predictable. There are "warm" colors, such as red, orange, and yellow, and there are "cool" ones such as green and blue. The architect has at his or her disposal a palette the use of which can have a direct impact on viewers' experience of visual images, and it has been used throughout history as a symbolic motif. Many buildings, such as the Gothic cathedrals, were brilliantly painted but have lost their color through weathering. Fortunately, the stained glass windows in these and other cathedrals have survived. It is widely understood that warm colors in a building are inviting and comforting, whereas cool colors have a distancing effect. As noted, color also has symbolic value, especially in non-Western religions like Buddhism (Ragsdale 2007).

Tresidder (2006) mentions the almost universal color symbolism of white and black. The former is associated with positive values and is the color for the clothing of brides and the confirmation gowns of infants. Black, by contrast, is negative, although it is a color often used by the clergy as well as in the academic regalia of university faculty members. Red suggests energy, gold implies royalty and divinity, green suggests renewal and hope, and purple implies dignity. Of course, each of these colors also symbolizes other qualities, and each signification may vary according to culture.

Closely related to color, in turn, is ornament. Architecturally, ornament is not merely decorative like the icing on a cake. Instead, it serves a communicative function. In Romanesque and Gothic cathedrals especially, the sculpture around entryways served to represent stories from the Bible and served as visual homilies (Ragsdale 2007). Gargoyles not only were to ward off evil spirits but were functional as downspouts. Ornament has been proclaimed both as essential to architectural design and as inimical to it, but this difference of opinion does not need to figure in any assessment of ornament as a feature of visual communication. As Strickland (2001, xiii) says succinctly, ornament is "to emphasize structure, embellish the surface with visual detail, model light and shadow, add human scale, and delight the eye." The latter is so important in the visual experience of architecture that one wonders what could possibly be the motive for a truly plain building.

Texture is the relative roughness or smoothness of a particular surface. As such, texture is related to both color and ornament. Texture, for example, is an important consideration in evaluating sculpture, with the highly polished surfaces of some bronze, silver, and brass pieces conveying a much different "feel" than the rather coarser textures of either marble or limestone. Roth (1993) distinguishes between optical and tactile texture. Even at a distance, some buildings will appear to be either rough or smooth. One of the desirable features of some Italian Renaissance buildings, such as the Palazzo Pitti in Florence, was a "rusticated" surface. "Rustication consists of courses of cut stone whose edges are chamfered (beveled), often at a 45-degree angle, and whose projecting faces are smooth or roughly textured and suggest solidity and mass" (Cole 2005, 272). The rough hewn stone appearance suggested the ramparts of a fortress and conveyed the idea of power and strength (Ragsdale 2009b). The titanium-clad surfaces of Frank Gehry's Guggenheim Museum Bilbao are highly light reflective and have rather a softer look than the rusticated surface. Up close, one may touch a surface and feel its texture. Again, smooth surfaces, such as furs, soft leathers, and rounded stones, are sensual and inviting. Coarser surfaces, such as wooden benches and concrete flower boxes, are cold and uninviting to the touch.

The acoustics of an enclosed space are another of the architect's concerns, and acoustics are affected to a large extent by such previously discussed elements as space and texture. The acoustics of a building are of greatest concern to the designer of performance halls, where the quality of a symphonic performance or the audibility of an actor's lines, depends on the shape of the space in which the performance takes place. Like light, however, sound within an enclosed space is pervasive and must be amplified, damped, or shaped in order to effect a pleasing experience for the visitor. There are more than 16,000 museums in the United States of America alone (http://www.greatmuseums.org), and interest in American museums appears to be at a record level. Museum visits are high on the list of tourist attractions in vacation spots, and an average of 2.3 million people a day visit an American museum. Given this large number of visitors to museums, it is crucial to design galleries both which will display the museum's collection advantageously and also help to offset the inevitable noises created by the shuffle of feet and the whispers of conversation. Roth (1993) acknowledges the importance of acoustics by devoting an entire chapter to the subject.

Strickland (2001) lists weight or mass as her final element of architectural design. Roth (1993) discusses "firmeness" instead. Both writers acknowledge the first of Vitruvius's three elements—strength—in their explanations. The first principle of any structural design, after all, is solidity, that is whether or not the building will stand up. Neither function nor beauty would be of any moment should the building collapse. With today's materials and engineering skill, one presumes on the solidity of a structure, but it has not always been so. The walls and spires of cathedrals, for example, have more than once fallen down, and the Campanile of Pisa is still leaning.

Weight or mass certainly is an important factor in the visual experience of a structure. There is the actual weight or mass as well as the apparent weight or mass. The Palazzo Pitti, for example, does indeed resemble a fortress, because of the solidity of its design as well as the rusticated texture of its walls. By contrast, the Ospedale degli Innocenti, also in Florence, presents a lighter, more airy, and inviting image to the visitor. Symbolically, the solidity of such museums as the British Museum and the Prado reinforces the notion of importance and dignity. The great mass of skyscrapers also reinforces their significance.

The second of Vitruvius's three essential features of architecture was utility or function, and Roth (1993) includes it in his list. Strickland (2001) does not, perhaps because it is such a broader concern than the others listed. To be sure, a building must be useful for the purpose of its design, nearly as much so as it must stand up. The architect will consequently make appropriate concessions to the client's needs, whether the structure be a personal residence, an office building, an opera house, an art museum, or a courthouse. Within the limits set by the necessity for strength and the desirability of utility, the architect is free to provide the viewer with visual delight or beauty. Beauty, it will be seen, is the Vitruvian element most closely related to the visual persuasiveness of a structure, but it comes only after the other two qualities have been perfected. The only caveat to this statement may be those comparatively few buildings whose beauty seems to arise out of their strength and utility. It may be, for example, that such a skyscraper as the Empire State Building in Manhattan is beautiful for the most part because of its strength and utility.

The final architectural element to consider here will undoubtedly strike the reader as odd. Roth (1993) alone lists ugliness. Such an element seems quite out of place, and it certainly would be were it to be thought of as a desirable attribute of a design. Roth acknowledges that there are several possible ways in which ugliness could be defined. Something without "a perceivable pattern of relationships," something "confusing because it is ambiguous," or something that "does not conform to accepted norms" is ugly (79). Roth recounts the familiar history of artistic styles in which "each generation . . . rejects its fathers and embraces its grandfathers." Renaissance writers, for example, disliked Medieval architecture, so they called earlier structures Gothic, "the work of barbaric Goths." In the Modern era, ornament was roundly discounted in favor of spare, simple work. What is ugly one day is attractive the next. "The value of ugliness is that it forces us to examine our accepted conventions; we may find that our preconceptions do not have much substance." These elements from Vitruvius, Strickland, and Roth provide a useful framework for assessing the visual experience of architecture, but it is necessary also to consider just what the visual experience of architecture means.

It would seem to be abundantly clear that architecture is communicative. Earlier it was suggested that it is also more often than not persuasive. Because visual persuasion is such an important concept in assessing the visual experience of architecture, it is appropriate here to consider in greater detail the case for structures as visual persuasion. Essentially, the question is whether or not when one looks at a building there is anything more in the experience than the messages that the structure is solid and the structure is functional. Further, one wants to know whether or not the form is visually compelling, whether or not it produces an aesthetic response in the viewer. If it does, it should not be hard to accept the premise that this response is made in reaction to a type of social influence.

Assessing architecture as visual persuasion requires one to ask the question "How does one determine the effectiveness of the design?" For a social scientist, the only real answer to the question would require measuring the effect. However, it is social science research itself which has amassed enough findings in the field of persuasion that it is possible to make an educated guess. The operative word, of course, is "educated." Political campaign advisers and advertising writers, among others, use this evidence to answer their clients' questions about how to succeed at their persuasive tasks. Teachers in communication studies advise their students about how to improve their effectiveness based on these scientific findings. Similarly, there is also research evidence permitting generalizations about how visual communication can be persuasive.

Like the methodology in previous studies of structures as visual persuasion (Ragsdale 2007, 2009a, 209b), the approach in the present work is analogous to advising a politician or an ad writer or teaching a student. It involves using well-supported theoretical generalizations to assess the *likely* effectiveness of the visual persuasiveness of architecture. This is not unlike the task of the traditional rhetorical critic, who draws on historical information about speeches and speakers to evaluate effectiveness. Before turning specifically to that task, however, it is necessary to provide a background for using the methodology described above. The next chapter, therefore, examines the elements of architectural design as they evolved over time in the appearance of specific buildings and architectural styles.

Some buildings are more obviously persuasive than others. Art museums, for example, are designed to compel the visitor to come and to set an appropriate tone or point of view for the visit. Cathedrals, especially Gothic ones, were seen as visual metaphors of Heaven both in terms of their symbolic cruciform floor plan and their light gathering windows. Their exteriors were also highly ornamented with sculptures of Biblical stories and teachings. Stained glass windows repeated this theme. Government buildings, such as the US Capitol and Westminster Palace, were meant to stand for the power of the state and to be formidable if not intimidating surrogates of national authority. Even at the level of county courthouses in the US, the symbolism of power and authority in the government's buildings may be clearly seen. The palaces of earlier eras had similar meanings and also illustrated the personal majesty of the ruler. Monuments proclaimed the might of the state and the glory of conquerors, or they memorialized great events of war and victory.

It is perhaps less easy to see how other structures are visually persuasive. Universities, however, are no less about social influence than they are about providing classrooms, and skyscrapers are statements of strength and influence as much as they are business locations. Performance halls are not merely venues, they are like museums in their intent to attract visitors and set appropriate tones. In their private residences, people have tried to make statements about themselves in comparison to their peers. Finally, even such quotidian structures as bridges, train stations, and airports have been built with as much concern for the visual impact on the visitor as for a specific need or location. One thing is missing before launching an assessment of a variety of structures as visual persuasion. That is a historical context or backdrop for considering that which is visually compelling.

CHAPTER TWO

ARCHITECTURE:
A BRIEF HISTORY OF ELEMENTS

There is no shortage of excellent histories of architecture and of architectural styles, therefore this chapter takes a somewhat different path. It is a fundamental thesis of this book that to understand the persuasive impact of architecture one must understand the role of beauty in design. It is through beauty that architecture has its primary persuasive impact on the viewer. The judgment of beauty is often a rather subjective act, but it is also a fundamental view of this book that the elements and communication strategies of visual literacy which underlie the elements and principles of architectural practice offer a means by which beauty can be evaluated with some objectivity.

There are several ways of studying the history of architecture. One way, for example, is to examine the development of structures country by country. Another is to treat architecture in a strict chronological way without regard to geographic location. Yet another is to study the evolution of architectural materials, such as wood, stones, cement, and metal, and certainly it is instructive to examine the development of architectural styles, such as the Classical, the Romanesque, the Gothic, and so on. Architectural styles differ from each other for the most part in their emphasis on elements such as vaults, domes, doorways, columns, roofs, and windows. Studying a style not only means identifying its distinctive elements but also affords an opportunity to establish the more fundamental elements of visual form underlying each element and to consider their visually persuasive impact. In this chapter, therefore, the evolution of architectural elements in terms of their contribution to the visual power of structures will be examined.

The story of architecture is, of course, the story of humankind. Primitive peoples, living before the advent of recorded history, sought shelter in caves and have left us with a record of their interest in art with drawings on the walls of caves in such places as Lascaux, France. Archaeological evidence suggests that our ancestors also built huts both large and small using wood and animal skins. At its most basic, a building consists of a floor, enclosing walls, and a ceiling. Like a cave or a hut, a simple frame building fundamentally provides shelter. Shelter, of course, is not enough, since there needs to be an entrance, a light source, and perhaps an opening of some kind to vent a fire or provide for air circulation.

For doorways as well as light sources, openings were made in the walls of buildings using very basic post and lintel construction. Briefly, two upright supports or posts were bridged by a horizontal member called a lintel. This construction is almost purely functional and, providing that the surrounding wall is sound, it is strong. It is well established that early peoples in Europe erected megaliths, such as the famous stone circles at Stonehenge in England and at Carnac in France (Strickland 2001). Although they are not properly speaking buildings, the megaliths at Stonehenge give us an early glimpse of the basic and enduring style of post and lintel construction for doorways.

A variation on the post and lintel system makes it possible to create windows. One can imagine, however, an early builder pondering his effort and wondering how it could be made more attractive. Whatever else we know about human beings, we know that there has always been an aesthetic impulse which affects behavior. "Remarkably early," Strickland (2001, 2) writes, "human beings reached beyond pure utility. Whether constructing in mud, wood, brick, stone, marble, or concrete, they erected shelters and shrines meant to impress." We actually do not have to guess about what such an early builder thought, since we have the historical record to guide us. We know, for example, that one of the first variations on the post and lintel system involved converting the posts to columns and then decorating them. Simultaneously, two other things occurred to early builders. The first was to use not just two columns but many, and the second was to vary the horizontal lintel by converting it into an arch. The use of repeated columns gave a visual rhythm to a building that was aesthetically pleasing. Moreover, replacing or surrounding walls with columns opened buildings up to light.

Egyptian and Mesopotamian Architecture

The Egyptians were fascinated by columns. Although best known in the history of architecture for the pyramids, the Egyptians built such structures as the Hypostyle Hall in the Temple of Amon at Karnak with sixteen rows of 134 columns (Strickland 2001). The columns were as massive as they were numerous. They were "12 feet thick and 69 feet tall"

(10). "Surfaces were covered with incised, painted hieroglyphics, like the tattoed man at a circus. . . . So ornate is the temple at Luxor that, when Napoleon's troops first spotted its ruins, the entire army, agape, halted spontaneously and grounded their arms to stare" (10-11).

There are many possible ways to account for the visual impact of these early as well as later columns. As noted, repeated columns gave visual rhythm to structures, they admitted light into the interior of a building, and they conveyed a sense of solidity and power, especially the huge Egyptian ones. At a height of 69 feet, columns surely conveyed thoughts of heavenly ascension as they soared above the ground level viewer. Also for the Egyptians, columns were probably analogous to such natural landscape items as trees and other plants. This seems so because of the sorts of plant-like decorations used on the columns in addition to the hieroglyphics. Capitals were rendered as palm leaves, lotus flowers, papyrus bulbs, and the like (Strickland 2001). Along with the paint used to decorate them, these features reminded the viewer of the beauties to be found in nature.

The Mesopotamian cultures of Assyria, Babylonia, and Sumer, lying between the Tigris and Euphrates Rivers, were responsible for the emergence of the city. These cultures, located primarily in what is today Iraq and Iran, are perhaps equally well-known in the history of architecture for massive tile-decorated gateways and pyramidal structures known as ziggurats. The Mesopotamians surrounded their cities with walls like fortresses and made the entrances into these places especially arresting visually. Although much of Mesopotamian architecture has been lost to time and deterioration, the Pergamonmuseum on Berlin's Museumsinsel features a reconstruction of the Ishtar Gate using brilliant blue tiles from the original. This four-story arched entryway was also decorated with images of gold dragons and white bulls.

Much later, the Mesopotamian civilizations of Persia and Assyria used columns like the Egyptians had in temples and palaces. Since construction techniques in the region utilized wood beams to support roofs, these columns were typically much less massive than Egyptian ones (Cole 2005). There were a variety of capitals. Double-headed bulls or horses were common at Persepolis in the Apadana Palace of Darius and his son Xerxes, as well as palm leaf designs and scrolled volutes resembling the Ionic order of Greek architecture. There were Greek craftsmen from Ionia working in Persepolis at this time (ca. sixth century BCE). Columns were typically fluted and had a variety of base styles, with bell-shaped ones being quite common.

The Classical Architectural Style

What is commonly referred to today as the Classical architectural style was begun in Greece, perhaps reaching a zenith in the city state of fifth century BCE Athens. However, Greek architecture, like Greek sculpture and Greek culture, was copied and improved upon by the Romans, so that much of what may be said about the Greeks applies equally to the Romans. Undoubtedly the most distinctive aspect of Greek, and later Roman, architecture was the combination of column, entablature, frieze, and pediment used in the construction of temples. This architectural feature was integral to the use of a pitched rather than a flat roof line. When subsequent generations sought to revive Classical ideals, whether in the Renaissance or later, this combination lay at the heart. The most famous, as well as perhaps the most perfect, example of this architectural style is the Parthenon on the Acropolis in Athens.

Fig. 2-1 Entablature, Pediment, and Doric Order, Bodleian Library

The roofs of the earliest buildings were flat, but a flat roof is notoriously difficult to waterproof. A pitched roof, a triangle viewed from each end, permitted easy rain runoff and at the same time invited a distinctive form of end decoration. Forming the triangle was first an entablature, which was the horizontal piece resting atop the capitals of successive columns. The entablature consisted of three parts: the architrave, the frieze, and the cornice.

The architrave was equivalent to the lintel of earlier post and lintel construction and ran along the tops of the column capitols. The frieze was the middle section of the entablature, and like the architrave it ran continuously above and the length of the architrave. Often there was a decorative band separating the architrave and the frieze. The frieze was also frequently decorated with a succession of triglyphs and metopes. Triglyphs were three vertical bands bunched together to symbolize the end piece of what formerly was a wooden beam (Vitruvius 1960). Between the triglyphs were the metopes, which were often decorated with bas relief figures.

The cornice was the covering projection above the frieze. In modern residential construction, the cornice is equivalent to the soffit and fascia board projecting outward from the exterior walls above which is the roof. The cornice was horizontal like the architrave and frieze and was decorated in a variety of ways. It formed the base of a triangular structure above the entablature called the pediment, which was topped with a raking cornice following the pitch of the roof. The triangular area thus formed within the pediment, which was often also decorated with sculpture, was called the tympanum.

Greek columns were of three principal types. These types were associated with orders or whole systems for designing columns, capitals, and entablatures to insure proper proportions. The orders were treated at length by the Roman writer Vitruvius, although the emphasis here will be only on the appearance of the columns themselves. The three principal types were the Doric, the Ionic, and the Corinthian, named after their geographical origins within Greece. The simplest in appearance, the Doric column, had a capital made up "of a cushionlike convex molding known as an *echinus*, and a square slab termed an abacus" (Cole 2005, 98). There was no base. The Doric order was "the preferred style of the Greek mainland and the western colonies (southern Italy and Sicily)" and "reached its peak in the mid-5th century BCE."

In Asia Minor and the Greek islands, the Ionic order developed along with the Doric and was also widely used by the mid-fifth century BCE. In subsequent revivals of Classical architectural style, especially in European and American art museums, the Ionic order seems to have been the most common. Visually, the most distinctive features of Ionic columns are that they are set on a base and that they have a double-spiral volute or helix as

Fig. 2-2 Ionic Order, Ashmolean Museum

their capital. "It appears to derive from naturalistic forms such as the nautilus shell or fern, and made an early appearance in so-called *Aeolic* capitals. The circular, central part of a volute is called the eye" (Cole 2005, 101).

The Corinthian order developed after the Doric and Ionic and was more ornate than either. In fact, it was not "a structural system" like the other two but was instead "purely decorative" (Cole 2005, 102). It was "an imitation of the slenderness of a maiden" (Vitruvius 1960, 104). The Corinthian order mimicked the Ionic, but its capital was taller than those of either the Doric or the Ionic orders and took the form of acanthus flowers and leaves. Corinthian columns adorn the entryway of the Pantheon in Rome. Although the columns certainly support the roof of the portico, they are not structurally integral to the building itself and in this sense are purely decorative.

Although not an order *per se,* another notable type of column from Greece was the caryatid, such as may be seen on the south porch of the Erechtheum on the Acropolis in Athens. A caryatid is a column consisting of the sculpture of a woman draped in cloth, with the woman's head serving

Fig. 2-3 Corinthian Order, The Circus, Bath, England

as the capital of the column. Not so widely used in Greece, perhaps because of the extra work required to sculpt the figure, there is an especially nice use of this type of column in the Louvre's Hall of the Caryatids. The female form also would have had specific connotations for the Greeks, as well as others, which would have been inappropriate in some temple settings. The Romans would later introduce two more orders. They were the Tuscan, resembling the Doric, and the Composite appearing to incorporate both the Ionic volute and the Corinthian ornamentation in its capital. The Romans would also use half-columns incorporated into walls primarily as an ornamental feature. These ornaments seemed to project from the wall and were called engaged columns (Roth 1993).. Another and similar Roman innovation was the pilaster, which was a flat wall projection decorated with the base and capital of a traditional column.

There is no dearth of examples of Classical architecture thanks to the many revivals of this style over the centuries, even though much of the Parthenon and other both Greek and Roman structures today lies in ruins. The British Museum, for example, is a design in the Classical manner. In the US, the Philadelphia Museum of Art is another remarkably attractive example of the style. So also are the Altes and Neues Museums in Berlin and the Madeleine in Paris.

The Egyptians and later the Romans also erected freestanding columns to serve as monuments. One notable such Egyptian column, the Obelisk of Luxor was confiscated by Bonaparte and removed to Paris where it stands today in the Place de la Concorde at the Louvre end of the Champs Élysées. In Rome, perhaps the most notable such monument is Trajan's Column. The French themselves built the Colonne de Vendôme, modeled after Trajan's Column, to commemorate Bonaparte's victory at Austerlitz in 1805.

Fig. 2-4 Hall of the Caryatids, Louvre

This discussion of the Classical style will not be complete without also noting several other aspects which have powerful visual effects. The temples and civic architecture of Greece and later Rome were very large. Their scale dwarfed lesser structures, including in Greece the rather simple personal residences of the citizens. The resulting mass and weight speak of great power and significance, and the visual effect is one of majesty. Especially in the case of the Athenian Parthenon, these huge structures were often constructed upon a hilltop site, or acropolis, that was equally compelling. Of course, such a site provided a defensible position from attack by enemies, but it also required viewers to look upward as if to the heavens. This later became a characteristic feature of art museums. The term "museum" refers to a temple for the Muses, and art museums have often been situated on hilltop or other similar locations (Ragsdale 2009a, 2009b). An elevated site is certainly not the only or even the most visually impressive location. Sometimes the site of a structure is compelling because it is centrally located or because it lies in proximity to another structure or structures of importance.

Returning to the Classical style itself, one must consider the distinctive features added by the Romans. Perhaps the most significant such feature was the arch. While arches had certainly existed before, the Romans perfected them. Roman arches were distinct in their structure. Constructed of stones over a wooden frame and finished with a wedged keystone at the top center of the arch, these structures were held together by the opposing force of each stone against its neighbor and finally against the keystone. The supporting wooden frame was removed once the arch was complete.

Freestanding triumphal arches were widely used as monuments to successful armies. The Arch of Titus in Rome is such a monument along with Bonaparte's Arc de Triomphe in Paris. The "original significance was religious. At the end of a military campaign, soldiers were encouraged to pass under an arched 'magic door' to siphon off their aggression" (Strickland 2001, 29). While the original purpose of an arch was to form an entryway, the Romans used this simple basic system to build much larger and more elaborate structures, such as aqueducts, vaults, and domes, throughout the Empire.

The Colosseum of Rome illustrates how the Romans used walls of arches one above the other to achieve great height and an imposing appearance. Aqueducts were constructed in the same way. The construction offered strength and solidity, but it was also visually compelling. Like an array of columns, the rhythmic repetition provided beauty to the structure. The Pont du Gard aqueduct, outside Nîmes in southern France, is a beautiful monument to the Roman arch. It also is a fine illustration of the integration of strength, utility, and beauty. That it stands today is evidence of its strength, and we know that it served to provide water for the Roman settlement at Nimes. Like the Colosseum, the Pont du Gard is built on a grand scale.

An arch extended along a horizontal line is a vault. Vaults were used for passageways and arcades. Vaults joined at right angles to each other created groin vaults. The Romans developed a form of concrete that permitted them to create structures not previously built. Concrete supported the vaults that were used for arcades and passageways. It must not have taken Roman builders long to realize that the use of concrete would permit widening their vaults enough to create rooms. The builders also undoubtedly noticed that the graceful curves of vaults naturally drew the eye upward, so that it was an easy transition from a vaulted room to a dome.

As Strickland (2001, 21) points out, "an arch, rotated 360^0, creates a dome." In Rome itself and throughout the Empire, massive interior spaces were created by the development of domes whose stones were, like those in Roman vaults, held together by concrete. Domes are especially significant visually. They not only allow for interior spaces of massive size, which was very useful for governmental meeting spaces, they are naturally adaptable to temples, basilicas, and cathedrals. In spite of the many similarities between Greek and Roman structures, the use of domes sets Roman architecture apart.

Perhaps the most visually impressive domed structure in Rome itself is the Pantheon, built by the Emperor Hadrian in the 1st century CE. This vast, single-roomed space was originally built as a generic temple in honor of all gods, and it has been used both as a governmental building and a church. The architecture combines the proportional structural designs of the rectangle and the circle. The building's floor plan is primarily a circle. However, there is a massive portico in the shape of a rectangle, which in terms of the floor plan overlaps the circle. This portico is supported by eight Corinthian columns along the front with two more on each side.

Byzantine Architecture

As the Roman Empire began to pass into history, the Emperor Constantine decided to protect his power by moving the capitol to present-day Istanbul in what is now Turkey. There he created a new city on the site of the old city of Byzantium and consecrated it in 330 CE. Constantine was the first Christian Roman Emperor, and the move to Byzantium/Constantinople marked three distinct architectural developments. Byzantine architecture was first and foremost religious, which was an important departure from the secularity of Rome. As religious architecture, it sought to represent Christian belief within as well as outside buildings. Third, it combined elements of Classical architecture with those of the Orient, especially those of Persia (Strickland 2001).

The three developments resulted in distinct floor plans, the use of pendentives as forms of support, domes, and mosaics and other forms of elaborate interior decoration. At first, floor plans were adaptations of the simple basilica. Basilicas were originally long rectangular governmental buildings and were chosen by Constantine as the model for churches. In the interiors of these basilicas were rows of columns, which set off narrow aisles on each long side of the larger space known as the nave. The nave was illuminated by clerestory windows above. This simple floor plan became the basis for the great Romanesque and Gothic cathedrals to follow.

A second significant Byzantine floor plan emerged in structures which "combined the basilica's horizontal axis with the vertical accent of a dome. They were frequently built on a Greek cross plan (with two arms of equal length crossing under a central dome) and could be round, octagonal, or square on the outside" (Strickland 2001, 33). The Romans built domes "over circular or polygonal spaces. Byzantine builders were able to place them over square or rectangular compartments by the use of pendentives" (Cole 2005, 312). Pendentives were inverted triangular vaults used for support so that Byzantine domes seemed simply to float in the air. This second floor plan characterizes the Hagia Sophia in Istanbul, the Cathedral of St. Basil the Blessed in Moscow, and, in the West, the Basilica of San Marco in Venice. Clearly the basic style varies with the geographic location.

Like the secular basilicas of Rome, Byzantine churches and cathedrals were brilliantly decorated within. Lavish decoration was intended to symbolize the Heavenly City (Strickland 2001, Ragsdale 2007). With ideas borrowed from both East and West, these Byzantine structures featured colorful glass and gold mosaics, inlaid marble, and sculpture. Later Romanesque and especially Gothic churches would elaborate on this theme with fine wood carving and large stained-glass windows. The windows served two functions: they told biblical stories and they permitted much greater levels of interior illumination to simulate Heavenly light.

Romanesque and Gothic Architectural Styles

The Byzantine style spanned more than a thousand years (330-1453 CE). Within this span of time, however, there were two perhaps even more significant architectural styles to develop: the Romanesque (1030-1200 CE) and the Gothic (1140-1500 CE). Like the Byzantine, these styles were primarily realized in religious architecture, giving rise throughout Western Europe to monumental churches and cathedrals (Strickland 2001). There had been virtually no architectural developments in the West for 500 years following the fall of Rome. Following the new millennium, and with the growing wealth of churchmen, there was a burst of building, and the principal structures were castles and churches.

At the beginning, these new structures were massive, solid, and followed the pattern of the architecture that the Romans had spread throughout Western Europe. Although it was not until the 19th century that they were called Romanesque, it is obvious why the name was appropriate (Strickland 2001). Romanesque churches and castles alike had thick walls with small windows to offer protection from marauding armies. They also used barrel vaulting and rounded arches. The churches were modified basilicas, in that they preserved the aisles, columns, and nave of the original governmental buildings but added a crossing section near one end. This section gave the floor plan a cruciform pattern imitating the crucifixion cross rather than the Greek cross of the Byzantine churches. This section of the structure was called the transept. The orientation of the floor plan was west to east, with the entrance and the narthex facing west. Ordinarily, there were three entrance portals, one each for the two aisles and the nave. The Gothic style, which supplanted the Romanesque, would maintain this cruciform floor plan, important for its symbolic recognition of Christ's cross.

The Romanesque style was important architecturally inasmuch as it represented a rebirth of building in Western Europe and reestablished the basic principles of Roman building. The building program itself was epic. "Between 1050 and 1350 in France alone, eighty cathedrals, five hundred large churches, and tens of thousands of parish churches were erected" (Strickland 2001, 38). In terms of visual persuasion, it is probably the scale of this architecture, its mass and size, that is most compelling. Within such buildings, although most were poorly illuminated, huge columns along each side of the nave and a vast vertical space suggested the grandeur of Heaven and pointed directly to it.

The Gothic style of architecture was a natural evolution from the Romanesque as suggested above. "Reliant on the sheer power of mass to abut and restrain the tremendous outward thrust of thick nave barrel vaults, Romanesque architecture could not open up to the light. . . . In the new phase into which medieval architecture now passed, the presence of light, the symbol of God's divine Grace, became the preeminent symbol; the church had to become transparent, and when it did so it was no longer Romanesque but Gothic" (Roth 1993, 288). Although it originated in France, Gothic architecture seems not to have had a particular name at the time. Only later was it referred to as Gothic, a term of disparagement used by Renaissance writers of the 15th century who were formulating new standards of taste (Scott 2003).

"It is possible to say that Gothic architecture was invented in 1141 for Suger, abbot of the monastery of Saint-Denis, a town just north of Paris" (Roth 1993, 291). Suger envisioned an opening up of the Romanesque style to more slender walls and, preeminently, to more light. "Increased verticality, a reduction in wall mass and the admittance of light through large windows filled with rich stained glass were the results" (Cole 2005, 200). Additionally, the Gothic style incorporated elements of the Romanesque which had not previously been used in the same structure: ribbed vaults, pointed arches, and flying buttresses. Although Romanesque buildings had utilized towers, it came to be a distinctive feature of Gothic exteriors that paired bell towers were erected over the west façade with spires over the crossing of the nave and the transept. Among other things, these towers and spires made the churches visible for miles around.

The Gothic style also utilized considerable external ornamentation, much more so than the Romanesque style had, and this evolved over the next three centuries into ever more elaborate decoration. Statuary, particularly along the western facades, and carved depictions of Biblical stories around the entrance portals were commonplace and taught the often illiterate visitor the ways of the church through "sermons in stone" (Ragsdale 2007). Similarly, the stained glass of the windows depicted more such stories. "The Gothic cathedrals were covered virtually from top to bottom with sculptural representation of biblical stories. The most dramatic innovation was the virtual elimination of the walls of the church; in their place appeared membranes of colored glass depicting stories from scripture" (Roth 1993, 293-294).

A thesis of this book is that architecture's visual impact derives primarily from its beauty, and its beauty in turn derives from the basic elements of architecture underlying it. In the Gothic style, more basic elements come together harmoniously than in perhaps no other previous way. These are scale, including mass and size, space, acoustics, light, and ornament. In addition, Gothic structures, especially cathedrals, were ordinarily centrally located in a city, making their site an additional basic element. No matter the misnomer—"Gothic"—which has survived, these structures are among the most compelling forms of any remaining today. Like the Classical style, Gothic architecture would also be reborn in the centuries following its reign as the dominant style.

Renaissance Architectural Style

The Renaissance brought with it changes in thought, writing, art, and architecture. Marked by the recovery of Greek and Roman artifacts and texts that were thought to have been lost, the Renaissance ushered in a new period of humanism, breaking away from the ecclesiastical domination of the medieval period. Although there is still a tendency to think of the Renaissance as a dramatic break from the past, modern scholarship has taught us that it was evolutionary. This was especially so in architecture. "Renaissance architecture in Italy is characterized by harmony, clarity, and strength. . . . [and] features the use of classical motifs and the architectural orders, or column styles, of antiquity" (Cole 2005, 228). However, this "did not signal a sudden break with Gothic or indeed earlier Romanesque styles."

Since the Renaissance began in Italy—in Florence to be precise—and since Florence remains primarily a Renaissance city, it is appropriate to look there for a synopsis of early Renaissance architecture (cf. Ragsdale 2009b). There is much significant architecture, including especially the Basilica di Santa Maria del Fiore with the Duomo of Brunelleschi, the Battistero di San Giovanni with Ghiberti's bronze doors, and Giotto's adjacent Campanile. North of the Duomo, there is the Ospedale degli Innocenti, which is probably the earliest Renaissance building. Just south of the Duomo is the Palazzo Vecchio, Florence's town hall, and the Ponte Vecchio, over which one can reach the Palazzo Pitti. Connected to the Palazzo Vecchio is the Galleria degli Uffizi. These are merely the most significant structures of what is surely an architectural gold mine.

Italian Renaissance architecture marked one major departure from the prevailing Romanesque and Gothic styles. Where these two architectural styles resulted in buildings with imposing mass and scale, the Italian Renaissance style most often resulted in structures of great delicacy and airiness. What were the most obvious features of the Italian Renaissance style? First, Florentine buildings reflected the ideals of Greek and Roman architecture. Basic geometric forms, such as the square, lay at the foundation of many new buildings. Proportion and symmetry were paramount.

Second, Florentine buildings often featured rhythmically repeated slender columns and semicircular arches. Filippo Brunelleschi's design for the loggia arcade of the Ospedale degli Innocenti (orphan's home) incorporates Corinthian columns with arches. Painted ceramic medallions by Luca della Robbia between the arches were added later. Such other notable buildings as the Doge's Palace in Venice, the Palazzo Medici-Riccardi in Florence, and the loggia connecting the two wings of the Uffizi follow Brunelleschi's lead.

Third, upper floors often repeated the semicircular arch theme with arched windows. Top floors featured an abundance of rectangular windows, which permitted higher levels of illumination than were possible below. Rooflines were typically not pitched and were brightly decorated with terra cotta tiles. Finally, the exteriors of many Italian Renaissance buildings were very plain although often painted in tones of ochre and yellow.

Fig. 2-5 Ospedale degli Innocenti

While these are the main features of Italian Renaissance architectural style to be found in Florence, there are other characteristics to be seen in churches and palazzi. The Convent of San Marco, with its frescoes by Fra Angelico, for example, used scrolled supports for its pediment. This gave the façade a triangular appearance. The triangle is the symbol of the doctrine of the Trinity, and it also suggests an upward thrust toward God and Heaven. The Pitti Palace does not appear to be even a part of the Florentine Renaissance style. However, in some buildings architects sought to give the impression of ruggedness and power through the use of "rusticated stonework" (Glancey 2006, 279). The Pitti Palace certainly is rugged in its appearance. That it was also a fortress perhaps explains this deviation from the more common patterns in Florence.

Of course, there is great variation within many architectural styles, and that of Renaissance architecture is no exception. Indeed, Florence perhaps best figures in the popular imagination for an altogether different but characteristic architectural element: the dome. The Basilica di Santa Maria del Fiore, in roughly the center of the old city, is a church in something of the Byzantine style with a dome designed by Filippo Brunelleschi. Most travelers know it as simply "the Duomo." It is easily the most visible structure on the Florentine skyline. The dome was 138 feet across and was inspired by Brunelleschi's study of the Pantheon in Rome. It was the largest dome to be erected after the days of the Romans and was itself the direct inspiration for Michelangelo's design for the dome of the Basilica of St. Peter in Rome (Strickland 2001).

Although Brunelleschi is revered as the father of Renaissance architecture in Florence, there is another architect whose legacy transcends even Brunelleschi's. He is Andrea Palladio (Palladio 1965). "His *Four Books of Architecture* (1570), which even Thomas Jefferson and George Washington owned, established his designs as universal prototypes throughout Europe and the Americas" (Strickland 2001, 65). Via Inigo Jones, an English devotee of Palladio, the Palladian style characterized architecture in England at least through the Georgian period and is therefore called Georgian. Palladio's greatest contribution to architectural style was his Roman-influenced emphasis on symmetry. His Villa Rotonda in Vicenza has Classical porticos on all four sides complete with Ionic columns and Greek pediments.

The Renaissance developed first in Italy and then migrated throughout Western Europe. As it did so, Renaissance architecture took on some of the characteristics of local, indigenous styles. The French king François I built his hunting lodge, the Château de Chambord, in the Loire Valley in the Palladian manner as well as the castle at Fontainebleau. He invited Italian artists and architects alike to work at his court (Cole 2005). The Renaissance style of Northern Europe mimicked that of the Italians in such buildings as the Town Hall of Antwerp and the Armory of Gdansk. The Escorial in Madrid shows the influence of the Italian architects, and of course the style was pervasive in England.

Inigo Jones's most famous example of the Palladian influence is undoubtedly the Banqueting House in London. Palladianism also characterizes many notable buildings in the US, such as Mount Vernon, George Washington's home in Virginia, Thomas Jefferson's home at Monticello, Virginia, and the University of Virginia in Charlottesville designed by Jefferson. In the US, the Palladian style is often referred to as Federal Colonial or simply Federal.

Baroque and Rococo Architecture

It is especially important, given the focus on visual persuasiveness in this book, to take note of the baroque style, which supplanted that of the

Renaissance beginning in Rome in the 17th century. "It was first and foremost an architecture of persuasion," writes Cole (2005, 256). "The act of stepping into a church became an experience in which symbolic and illusionist schemes appealed as much to the emotions as to the intellect of the faithful." As noted earlier, the baroque was not the first such obviously persuasive style, with the Gothic having been equally overt in its intent to influence. The preeminent example of Roman baroque style is the Basilica of St. Peter.

The term baroque figures in the popular mind as one indicating an especially ornate style. Indeed, "it is characterized by exuberant decoration, expansive curvaceous forms, a sense of mass, a delight in large-scale and sweeping vistas, and a preference for spatially complex compositions" (Fleming, Honour, and Pevsner 1991, 31-32). Many well-known examples are to be found in Paris, including the east façade of the Louvre, the Palais du Luxembourg, and Les Invalides. The paramount example is surely the Château de Versailles outside of Paris. In London, it is Christopher Wren's design for St. Paul's Cathedral. In Stockholm, it is the Royal Palace.

By the early 18th century, French architects, influenced by Louis XIV's extravagance at Versailles began to develop an even more ornamented style than that of the baroque called rococo. As noted in Chapter One, ornamentation is an important element in architecture and one which for the most part was not merely a cosmetic afterthought but an integral aspect of structures. Rococo style, however, sits perilously close to the edge of mere decoration and may be responsible in part for the radical rejection of ornamentation by the architects of the 20th century. "It was characterized by white and pale pastel colors, lavishly accented with gilding. Light in color and light in feel, the wispy décor featured plaster flowers, shells, scrolls, clouds, and lacy tendrils" (Strickland 2001, 74).

With the rococo variation on the baroque, "architecture became, quite literally, an exquisite and colorful veneer that was applied over something else; it was visual effect with very little structural truth" (Roth 1993, 394). The term baroque came from a "derogatory Portuguese term used for misshapen pearls: *barocco*" (357). Rococo came "from the French *rocaille,* used to describe the shell-encrusted grottos fashionable in gardens at the beginning of the eighteenth century" (389). Both words were terms of disparagement in the vocabularies of the 19th century Neoclassicists.

Classical and Gothic Revival

By the middle of the 18th century, certainly by the 19th, and especially in France and England, a reaction to the baroque and rococo styles resulted in another type of renaissance, this one a revisiting of the purely Classical styles of Greece and Rome. Called Neoclassicism, this rebirth gave the world the Paris Pantheon, London's British Museum and St. Pancras Church, and in the US the Capitol in Washington, DC, as well as many state capitol buildings and an impressive number of major art museums such as the one in Philadelphia. The Brandenburg Gate in Berlin is a Neoclassical structure, along with the Neues Museum and Munich's Glyptothek.

This period also saw another revival, this time a return to the Gothic architectural style. This "corresponded to the more romantic side of eclecticism; for just as the trabeated Neoclassical orders suggested enlightened logic and ennobling probity, so craggy and dark Gothic architecture corresponded to the Romanticists' desire for mystery and irregularity of form" (Roth 1993, 426). The preeminent example of the Gothic revival in England is the Palace of Westminster in London. The House of Parliament in Budapest, Hungary, is Gothic revival as is the National Cathedral in Washington, DC.

It would certainly be an oversimplification to regard the development of architectural elements and styles as a consistently expanding spiral, although the temptation to do so is strong. However, the opening of the 20th century brought such a radical shift in the philosophy of architecture, and to be sure the world of art in general, that it is nearly impossible to see the shift as anything but a complete disjunction from what had come before it. The only way to see modernism and its relations as part of what had gone before may be to think of it as what remains when all but strength and utility are left of the Vitruvian triad. In this respect, the 20th century probably is best characterized as an era in which architecture was almost completely transformed.

Modern Architectural Styles

The 20th century is replete with "isms:" modernism, post-modernism, brutalism, Neomodernism, deconstructivism, formalism, and urbanism (Strickland 2001). It is also the century of the Bauhaus and of International Style. Where in the previous two centuries architecture had become more and more ornamented, in the 20th it became less and less so. "Form follows function," proclaimed Louis Sullivan (Strickland 2001), and predominantly that meant simple geometric forms, straight lines, and few or no decorative elements. As in the case of 20th century art, architecture became more and more difficult to describe as a visually compelling form, especially in the sense of beauty leading to persuasiveness. Modernism is probably the best place to begin to examine the contribution of this century.

As Strickland (2001, 118) notes about the first half of the 20th century, "architects tried to change the world, remaking it in a stark image of abstract geometry." The Bauhaus or International Style of this period emphasized "unpainted, unmolded, unadorned modern buildings" where "standardization, function, and structure reigned supreme" (120). The Bauhaus was a school for architects established in Weimar, Germany, in 1919 and led by one of the century's preeminent architects/thinkers: Walter Gropius. The Bauhaus principles evolved into what has been termed the International Style, represented in the US by such architects as Philip Johnson and in Germany by Ludwig Mies van der Rohe, he of "less is more" fame. It "is characterized by asymmetrical compositions, unrelievedly cubic general shapes, metal and glass framework often resulting in large windows in horizontal bands, an absence of mouldings, a predilection for white rendering and open ground plans, loosely arranged" (Fleming, Honour, and Pevsner 1991, 227-228). Mies and Johnson designed the Seagram Building in Manhattan, which serves as a striking example of the International Style.

The 20th century was not only one of straight lines, sharp angles, and unadorned structures. It was also the century of Frank Lloyd Wright and Le Corbusier to mention only two of the many prominent architects of the time. Wright's designs for the Robie House in Chicago and Fallingwater in Bear Run, Pennsylvania, were angular enough, but by the end of his long career the master's Guggenheim Museum in Manhattan sported the circular lines of an inverted ziggurat (Ragsdale 2009a). Le Corbusier (Charles-Édouard Jeanneret), like Wright, designed buildings such as the Unité d'Habitation in Marseilles, France, which were models of modernism, but he is probably best remembered for the design of Notre-Dame du Haut in Ronchamp, France. This church is "eminently expressive, with its silo-like tower, [and] its brown concrete roof like the top of a mushroom" (Fleming, Honour, and Pevsner 1991, 263).

While this account is merely suggestive of 20th century architectural styles, it is sufficient to identify the most challenging issues with which one interested in visual persuasion must work. Absent traditional structures of beauty, such as columns, arches, domes, sculpture, and in many cases color, how may 20th century buildings be thought to have beauty? It would, of course, be possible to take the position that the straight lines and sharp angles are themselves beautiful. From the standpoint of the elements and communication techniques of visual literacy, a rather strong case may be made for this position. For those who prefer the spare lines of Swedish and Danish furniture of the 20th century, as opposed to the ornate furniture of Louis XIV, this case would be readily apparent. On the other hand, one might simply take the position that 20th century style is simply neither beautiful nor visually compelling. Roth (1993) comes rather close to this view.

Like modern art (Ragsdale 2009a), modern architecture breaks so radically with the past that the usual standards for beauty are difficult to apply. The International Style changed the standard for beauty by equating function and beauty. "Ambiguity, surprise, and delight in architecture were disregarded as irrelevant" (Roth 1993, 481). Roth says that "modernism was based on a number of fundamental fallacies" with "the most pervasive" being "that function in architecture was simple and therefore as easily analyzed as in a linear, industrial manufacturing process."

The most productive response to the question of how to evaluate the beauty of 20th century architecture, and thus its visual persuasiveness, is that in spite of the modernist equation of function and beauty the visual impact of these structures will necessarily have to be assessed in terms of their architectural elements. As Chapter One demonstrated, the basic elements of visual literacy did not change in the 20th century. Indeed, beginning with these elements may afford more insight into modern design than any other approach. The communication techniques used with these elements also did not change. It is one of the challenges of the remainder of this book to assess 20th century work in its own terms from the standpoint of its most basic elements and techniques.

CHAPTER THREE

TEMPLES OF THE MUSES

Museums, especially museums of art, are an especially rich source for a first assessment of architecture as compelling form. Museums have previously been studied extensively for this purpose (Ragsdale, 2007, 2009a, 2009b), which means that parameters of judgment have already been established such that the task at hand may be eased. Additionally, the previous work means that new museums may be considered that were previously excluded. That museums are visually persuasive may at one point recently have been a novel idea, but even without the work by the author mentioned herein there is now no doubt that these structures not only are persuasive but are seen to be so by many different kinds of authors.

Macdonald (1996, 14), for example, writes that "any museum or exhibition is, in effect, a statement of position. It is a theory: a suggested way of seeing the world." Hein (2006, xix) says that "the museum is, in a sense, imperceptible; it is the invisible eminence that wields power to move and reassimilate us." She adds that museums have "the unique ability . . . to change the way people think and feel, affected indirectly through the manipulation of things" (xxi). "The very establishment of a museum is a persuasive act" (Ragsdale, 2009a, 1).

As means of visual persuasion, museums may be assigned to one or more of five categories. In *Structures as Argument* (Ragsdale 2007), the author developed a five-part typology of museums based on persuasive intent:

- The museum as cultural icon
- The museum as polemic
- The museum as collective memory
- The museum as partisan advocate
- The museum as pure visual persuasion

This typology was used in whole or in part in two subsequent assessments of American and Western European museums as visual persuasion (Ragsdale 2009a, 2009b). It is quite straightforward.

Museums which are cultural icons raise the status of a nation or place by embodying its ideals. Such museums are the most common of the five types and include the Louvre, the British Museum, the Prado, and the Rijksmuseum. In the US, they include the Metropolitan Museum of Art, Chicago's Art Institute, and the National Gallery in Washington, DC. Such museums as these intentionally glorify their countries or their cities with their architectural splendor and their holdings.

Polemic museums embody extreme statements of appeal, such as the outcry against the Holocaust seen at Dachau and at the US Holocaust Memorial Museum in Washington, DC. The House of Terror in Budapest represents outcries against both Nazism and Communism. Both Nazis and Communists used the building which is now the museum. Museums sustain collective memory when they memorialize an event or person so that no one may forget what happened. The Anne Frank House in Amsterdam is such a museum. Museums as polemic and as collective memory are similar, with the former being strident, while the latter is more matter-of-fact.

Art movements and artists have sometimes been unaccepted by their contemporaries. Some continue to be, as is the case with much of the work of the 20th century. There are many museums whose primary role is to insure the acceptability of such work, and they do so by housing it and promoting it. Museums devoted to unpopular or misunderstood art are museums of partisan advocacy. Today many of the most important museums in the world are museums of partisan advocacy, including the Museum of Modern Art in Manhattan and the Tate Modern in London.

Most of the world's museums are visually impressive, but a few stand out for being so visually compelling. These are the ones which may be called museums of pure visual persuasion. The Louvre and the Centre Georges Pompidou in Paris, the British Museum, and the Prado are all architectural marvels in their own right. In the United States, this category includes the Metropolitan Museum of Art, the Philadelphia Museum of Art, the Guggenheim Manhattan, and the like.

These categories were not ends in themselves but served the purpose of putting together those museums whose visual persuasiveness was similar and internally consistent. The typology was also not intended to be exhaustive. For example, it omits museums of natural history, examined in *Structures as Argument* (Ragsdale 2007). Indexical museums, that is to say collections of actual objects such as aircraft and firearms, also do not fit.

The focus in this book and consequently in this chapter will necessarily be somewhat different than that in previous work. Here the focus will be on the last of the five categories mentioned above: the museum as pure visual persuasion. As will be the case throughout this book, the first question to ask is "How are museums beautiful?" Following immediately, the second question is "How do museums persuade?"

Before proceeding even to a review of the findings about specific American and Western European Museums, it is possible to offer a general multi-part answer to the question of how museums function as persuasive entities. If one thinks, for example, of the idea of the museum as pure visual persuasion, then one could immediately say that museums serve to draw visitors, to compel attention to this wondrous sight. Who would not want to visit the Getty Center in Los Angeles even apart from its contents? Beautiful buildings also imply great wealth on the part of their creators, and certainly the Getty fulfills that role. In this context, such structures convey to their locales a considerable amount of cultural status. The Louvre comes to mind immediately in this context. What would Paris be culturally without the Louvre? Then there are those museums whose primary persuasive function is to convince visitors and the public at large that their contents are valuable because they are housed in a museum. Even those who appreciate so-called modern art surely recognize that many items even at the Museum of Modern Art in Manhattan would not be noticed, not to say admired, apart from their display there.

The first two sections of this chapter will review the findings of previous studies of specific American and of specific Western European museums with respect to the questions mentioned above. Then the third section will examine a type of museum not previously considered as a form of visual persuasion. This "new" museum might be called the omni-museum, inasmuch as it contains within its walls—and outside them as well—virtually all of the works of art, cultural artifacts, decorative items, furniture, and even natural history items that are usually parceled out to separate museums. Of course, some present-day art museums, such as the British Museum for example, were at one time omni-museums, and such museums as the Victoria and Albert in London even now come close to the all-inclusiveness suggested here. Thus, the omni-museums to be considered here are not unique so much as they are unusual. Moreover, they are more likely than not to be found in areas of the world without an ancient cultural tradition and without easy access to the usual avenues of aesthetic commerce. In fact, the museums to be considered here are all to be found in the Southern Hemisphere, specifically in Australia and New

Zealand. First however, what about the pure visual persuasiveness of American and Western European museums?

One certainly could make a case for the pure visual persuasiveness of virtually all American and Western European museums. After all, it was the presumption from the outset that the home of great art ought also to be great. "Art museums have always been compared to older ceremonial monuments such as palaces or temples. Indeed, from the eighteenth century through the mid-twentieth centuries, they were deliberately designed to resemble them" (Duncan 1995, 7). "The central domes of older museums allude to the temple of the muses" and "walking up the entrance stairs elevates you out of ordinary reality into the art world" (Carrier 2006, 19). Art museums were intended to be compelling visual forms, and it is a rare one indeed that is not. However, there are so many museums in the US and Western Europe that it is helpful to single out a few which are notable for their architectural significance.

American Museums of Pure Visual Persuasion

In America, the following museums are notable examples of the impact of architecture as visual persuasion:

- The Metropolitan Museum of Art
- The Philadelphia Museum of Art
- The Kimbell Art Museum

Of course, other writers might create different lists, but this group of three has already been distinguished for its architecture (Ragsdale 2009a). Two of the three are traditional architecturally, while the third is a modern design with reflections of Roman architecture.

The Metropolitan Museum of Art is America's greatest museum, and its architecture is perfectly consistent with greatness. The building is imposing as is its location. The structure occupies four city blocks along Fifth Avenue, a prime area of real estate just north of midtown. This location argues for the value and importance of this institution to the city. The argument also applies to the nation as a whole. The US is not commonly thought to be a center of sophistication in the arts, so the presence of the Met is a compelling counterargument to that view. Additionally, the building is an architectural gem.

Fig. 3-1 Metropolitan Museum of Art

The architecture is Neoclassical, although not purely so. The building does not feature the signature Greek pediment above its columns, but it does make use of Classical design elements. The entrance is flanked with two pairs of Corinthian columns which rise two stories. The doorway uses two one story Ionic ones, and this motif is repeated in a regular succession all along the front of the museum. Above these smaller columns are arched windows. Carved medallions appear to the upper left and right of these arches somewhat reminiscent of Florence's Ospedale degli Innocenti. From the center out to the left and right, the façade of the Met is beautifully symmetrical. The Met is easily the equal of any building in Manhattan in the beauty of its architecture.

The size of the building becomes even more apparent as one ascends the marble stairs. Once inside the entrance rotunda, with its two-story ceilings, massive arches, and brilliant window illumination, a sense of awe is unmistakable. While this same sense can be experienced in many of the world's culturally iconic museums, there is none more awe-inspiring or beautiful. Site, size, ornamentation, rhythm, and proportion combine in the architecture of the Met to produce a visually compelling form, which certainly must account for some of the motives for visitors to come there. Apart from its collection, which is unmatched among museums after the Louvre, the Met is a powerful means of visual persuasion.

More than any other museum in the US, it is probably the Philadelphia Museum of Art which most obviously epitomizes the idea of a temple of the muses. Its site is commanding in many ways. The museum is built on a granite hill, which can be seen from wherever in the city the view is not blocked by skyscrapers. This site recalls the hill of the Acropolis in Athens, and the visual argument is that what is housed here has great significance for this city. This significance is accentuated by the stairs ascending from street level. The museum is at the northwest end of the Benjamin Franklin Parkway, a tree-lined avenue whose southeastern terminus is the Philadelphia City Hall. Though not quite so grand, this avenue mimics Paris's Champs Ėlysees. Behind the museum one finds the largest city park in the US (Babbitt 1995). In front and at the head of the entry stairs, there is a massive octagonal bronze fountain at the east entrance. In the plaza across from the southeast entrance, there is a very large statue of George Washington. So not one but many visually persuasive symbols combine in the Philadelphia Museum of Art to make it the architectural masterpiece that it is and a lure for visitors. Moreover, the building is so dramatic that it undoubtedly confers status on the art within.

Fig. 3-2 Philadelphia Museum of Art (under renovation)

The architecture is traditionally Neoclassical featuring multiple examples of the distinctive Greek pediment. It is described on the museum's website as "echoing the design of a Greek temple but of more massive Roman proportions" (http://www.philamuseum.org/information). Its exterior is honey-colored Minnesota dolomite. The main entrance is formed of eight Ionic columns and pediment but with an unadorned tympanum. North and south wings extend symmetrically out from the central part of the building providing a perfect proportion.

There is an especially pleasing aesthetic feature to be found on the north wing of the museum. The tympanum above the wing's columns consists of ten sculpted figures of Greek gods and goddesses of sacred and profane love. Like many Classical sculptures were, these figures are painted bright colors. Unlike the Classical ones, however, these statues have a ceramic glaze intended to keep the colors from fading. The tympanum is quite imposing. Seventy feet wide at the base, it rises to twelve feet in height at its apex. The Philadelphia Museum of Art is also a structure of huge scale. Weight and mass are likewise impressive. The building's honey color is especially appealing at sunrise and sunset, and the north wing tympanum adds to the effect of color.

Fig. 3-3 Philadelphia Museum of Art, North Wing Tympanum

The most compelling architectural feature of the interior of the museum is the Great Stair Hall at the east entrance. A wide staircase, without railings, repeats the symbolism of ascension to the realm of the muses. At the head of the stairs, is a replica of Saint-Gaudens's statue of the goddess *Diana,* while over the stairwell itself is Alexander Calder's white assemblage called *Ghost.* On the second floor balconies, there are the Constantine tapestries of Peter Paul Rubens. The stairs, statue, mobile, and tapestries combine to convey an impression of admission into the world of art and to prepare the visitor for what is to come, like an overture to an opera.

Fig. 3-4 Kimbell Art Museum

Just west of downtown Fort Worth, Texas, one will find a trio of museums, all of which are modern architectural designs. All are beautiful examples of contemporary style, although each is quite different from the others. There is the Kimbell Art Museum, the Amon Carter Museum and the Modern Art Museum of Fort Worth, all within a short walk of each other. All command a clear view of the city, and the Will Rogers Memorial Auditorium and Coliseum is nearby. The Kimbell is in many respects the most distinctive of the three, which is the reason for its treatment here. The architect Louis Kahn designed a unique modern structure which nonetheless evoked design features as old as ancient Rome and the baths of Caracalla. The building is a series of sixteen connected barrel vaults of poured concrete trimmed with Italian travertine. Although the building is not notably massive, this use of barrel vaulting suggests a pleasing solidity. The resulting repeated curves are aesthetically pleasing.

The interior of the Kimbell is one of the most beautiful places to display art in the world. The source of this beauty is the design for lighting. Each vault has a clear plastic skylight. Translucent reflectors were installed within and directly below the skylights to diffuse sunlight and to divert it along the walls of the galleries. The lighting enables the viewer to see art in the most pleasant of environments.

These three examples are only fractionally representative of America's many museums. Another writer might easily have chosen three completely different ones to make the claim that art museums, both traditional and modern, are often choice examples of architecture as visual persuasion. The Getty Center in Los Angeles, the Getty Villa in Malibu, the National

Gallery in Washington, DC, the Boston Museum of Fine Arts, the Museum of Contemporary Art in Chicago, and certainly the Museum of Modern Art in Manhattan could equally well exemplify this claim. Of course, museums in Western Europe offer another set of warrants for the claim.

Western European Museums of Pure Visual Persuasion

In Western Europe, the following museums provide examples of the impact of architecture as visual persuasion comparable to those in America:

- The Louvre
- The British Museum
- The Tate Modern

Again, other writers might provide different lists, but the intention here as before is to be illustrative. These three museums have also already been distinguished for their architecture (Ragsdale 2009b).

Fig. 3-5 Louvre Pyramid

The Louvre may well be the most beautiful of all museums. The evidence begins outside with the site (Strickland 2001). Originally designed as a fortress, the Louvre lies on the right bank of the Seine in the heart of Paris. The fortress eventually became a pair of royal palaces, one facing east and the other west, connected by enclosed corridors on the north and the south. The Tuileries palace on the west side was destroyed by fire and never rebuilt. In the place of the former palace is the Jardin des Tuileries. To the west in a straight line are the Place de la Concorde, the Champs Élysées, the Arc de Triomphe, and la Grande Arche de la Défense. There is no more majestic layout in the world.

The scale of the Louvre is massive. It was, after all, both a fortress and a palace. The word "overwhelming" is often encountered in discussions of a visit to the Louvre. Light is not only an important ingredient in beautiful architecture, it is a crucial element of installation design, especially for paintings, but the most important aspect of light in the Louvre today is the steel and glass pyramid of I. M. Pei in the Cour Napoléon. In addition to the main pyramid, there are three smaller ones and another which is inverted, all of which admit light into ordinarily dark areas underground. The great mass and weight of the older buildings is lightened by these airy structures, which are further enhanced in their luminosity by reflecting pools adjacent to the main pyramid.

Fig. 3-6 British Museum

From the standpoint of architectural style, the Louvre is predominantly French Classical Baroque (Howells 2002). The present complex evolved over many years, thus there are several other stylistic themes. The 16th century façade of the Cour Carée, for example, was designed by Pierre Lescot. Louis Le Vau, who designed Versailles, also worked on the complex along with Claude Perrault. Nevertheless, the Louvre is uniform in its appearance. The Cour Carée has three aboveground stories. The lower two are of normal height, while the third is an attic story. The elements of site, scale, and light make for a building complex that is pure visual persuasion. There is no denying the splendor of the Louvre, but that splendor comes from the Louvre's having been a royal palace. The present British Museum, on the other hand, was created specifically to be a museum, and it succeeds very well at its purpose.

As Wilson (1989, 7) puts it, "The British Museum is not only an international institution, it is a building. Its great Greek Revival façade is almost the symbol of a museum. . . . The entrance through a massive portal, past strong iron gates, instills a sense of awe and security." The Great Russell Street entrance to the museum is fronted by a great open square, with a stone staircase stretching completely across the entrance. The façade is a majestic double row of Ionic columns topped by a pediment of figures depicting the emergence of mankind from pre-civilization (Wilson 1989). The originally open Great Court within took on a new artistic beauty when a domed skylight was added in 2000.

The weight and mass of the British Museum are perhaps the building's most notable features architecturally, although its scale does not compare to that of the Louvre. Two wings extend outward from each side of the central part of the structure, which is essentially a large square with two floors above ground. The Ionic columnar design is repeated within the Great Court giving a pleasing sense of rhythm and of repetition. Within the building, the upper galleries as well as the single story Parthenon Galleries use skylights to provide illumination. All of the galleries are spacious and have high ceilings, which features facilitate the display of the very large items.

The site of the British Museum is not a hilltop, like the Parthenon, nor is it beside a river, like the Louvre, nor does it stand at the end of a grand avenue, also like the Louvre. Instead, it is in a central location in the Bloomsbury section of London somewhat like the location of the Met in Manhattan. The site is a bit north of Covent Garden, home of the Royal Opera, in a residential area interspersed with parks and squares. This area and the museum as well are like an oasis, a place where time and circumstance have fallen away. The location and the building itself have a visually persuasive power, and the building structure is very much in harmony with its contents. It magnifies the collection and stands as London's principal cultural icon.

There are those who find the contemporary museum architecture of the Centre Georges Pompidou in Paris shocking. Such people are surely also likely to be surprised by the Tate Modern at Bankside. The building was formerly a huge electrical power plant, which was abandoned in 1982 and scheduled for demolition (http://www.tate.org.uk). However, in a thorough modernization of the south bank of the Thames, the power plant was kept and remodeled into a center for modern and contemporary art. It opened in 2000 along with the London Eye and the Millennium Bridge. Like the art within, the Tate Modern is essentially spare, nontraditional as a museum, and an exemple of found art.

As visually compelling architecture, the Tate Modern's primary strength lies in its site along the south bank of the Thames in the shadow of St. Paul's Cathedral. Next door is the recreation of Shakespeare's Globe Theatre. The building's mass and weight are impressive. The Swiss architecture firm of Herzog and de Meuron left the exterior of the power plant largely intact. The building is seven stories high on the river side and retains a chimney from the power plant which projects considerably higher. At the top of the chimney is the Swiss Light, designed by the artist Michael Craig-Martin (http://www.tate.org.uk), which at night serves as a beacon over London. The architects stripped the interior bare, which has left cavernous spaces on each floor. The former Turbine Room is especially cavernous. It stretches from the east to the west side of the building and is also open to the ceiling.

The architectural style of the structure is a simple rectangle. The walls are flat, and there is no adornment. Both the previous generating station and the present museum are fine examples of the twentieth century International style of architecture, conceived and practiced by such men as Walter Gropius, Mies van der Rohe, Le Corbusier, and Frank Lloyd Wright. Transforming the power plant into a modern and contemporary art museum turns out to have been very successful. "It was recently voted the capital's favourite building," and "In five years more than twenty million visitors have taken possession of the building itself, notably enjoying the experience of being in the great Turbine Hall (Gayford, Holden, Moore, Smith, Snow, and Travers 2005, 5).

Fig. 3-7 Tate Modern

The spare exterior is matched within the building. The floors are bare wood, and the exhibition spaces are merely white cubes (Ragsdale 2009a). This style is not only consistent with other modern and contemporary art museums, it is consistent with the art of the twentieth and twenty-first centuries. Moreover, "Tate Modern is one of the few buildings . . . that take your breath away—especially when you walk into the Turbine Hall for the first time. It still holds a fascination because of the sheer scale of it" (Gayford et al. 2005, 19). The Tate Trustees claimed by 2005 that the Tate Modern was "the most popular museum of modern art in the world" and "the third most popular free visitor attraction in London" (41). As a piece of visual persuasion, the Tate Modern is something of a *tour de force.*

The American and Western European museums discussed here illustrate common themes among visually persuasive structures. The Met, the Philadelphia Museum, the Louvre, and the British Museum are all cultural icons. They lend status to their locations. The Kimbell and the Tate Modern, by contrast, are institutions of partisan advocacy, which provide imprimatur for the works of art they contain. What each of these also do is serve as compelling forms to draw visitors and to have social influence. Preeminently, these museums are prime examples of the museum as pure visual persuasion. They are compelling works of art in and of themselves and remind us of the power of the elements of architecture to form works of beauty and influence. Newer and somewhat different kinds of museums in Australia and New Zealand offer the opportunity to see whether or not such conclusions can be extended.

The Omni-Museums of Australia and New Zealand as Pure Visual Persuasion

One of the unfortunate side effects of the proliferation of museums, especially in the US and Western Europe, is that each new museum has fewer and fewer art objects available to form a collection. The total number of museum quality pieces of art is quite finite, and the number of great pieces still available on the market has hovered near zero for perhaps one hundred years. Australia and New Zealand, for all their natural beauty, fine wine, and burgeoning film industries, suffer inordinately from these side effects because of their comparative youth as advanced civilizations. The adverse effect of their youth is exacerbated even today by their relative remoteness from traditional centers of art. However, the same impulses which led to the creation of the Met, the Art Institute, and other American museums are to be found in Australia and New Zealand. The aesthetic impulse is a human one that is not bound by either the age of a culture or its geographical location. As a result, the visitor to these two Southern Hemisphere countries will be very gratified indeed by the omni-museums to be found in Melbourne, Australia, and in Wellington, New Zealand. Both traditional and contemporary art museums can be found in Sydney, Australia, and there is a very fine traditional one in Christchurch, New Zealand. While not omni-museums, the Art Gallery of New South Wales and the Museum of Contemporary Art in Sydney, as well as the Christchurch Art Gallery, will nonetheless be considered to give a complete account of art museums in Australia and New Zealand.

Two institutions in Melbourne are striking examples of what has been termed the omni-museum. These are the Melbourne Museum and the National Gallery of Victoria. Perhaps even more striking is the Museum of New Zealand in Wellington, most commonly referred to by its Maori name Te Papa Tongarewa, which means "container of treasures" (www.tepapa.govt.nz/AboutUs). Of the three, the National Gallery of Victoria comes closest to the concept of a culturally iconic museum, whereas the Melbourne Museum and Te Papa Tongarewa are outright and unabashed omni-museums. All are quite impressive examples of 20th

century architecture. Since the previous museums in this chapter have already been studied in more depth than just their architecture (Ragsdale 2009a, 2009b), these three museums will be discussed here more fully than the American and Western European museums.

Fig 3-8 Melbourne Museum

Located in Carlton Gardens, a large green park north of Melbourne's commercial district, the Melbourne Museum is a very new and very contemporary structure which is as compelling as modern architecture can be. Although the museum was established in 1854, the present building dates to 2001. It faces across a large public square the 19th century Royal Exhibition Building, which is a beautiful structure in its own right. The Melbourne Museum is the Exhibition Building's architectural counterpart. It features a two-story façade which extending horizontally across from the Exhibition Building. Perfectly symmetrical, this façade is all glass and metal. It has a hard texture, but it is sleek and catches and reflects light so as to give an open impression. Scaffolding runs along the front of the building and supports overhead canopies. The same openness continues within, with the entrance hall open to the upper galleries. Like other museums of pure visual persuasion, the Melbourne Museum draws visitors regardless of its contents.

The contents of the Melbourne Museum run the omni-museum gamut. There is an exhibit focusing on the Depression-era racehorse Phar Lap, including the preserved horse himself. There is also a full skeleton of a blue whale, the largest mammal in the world. Insects, fish, birds, snakes, and the like are also here. There is aboriginal art and a 150 foot long zinc etching at the entrance by the aboriginal artist Judy Watson. The upper level of the museum has a Mind and Body Gallery, a Forest Gallery, and a Melbourne Gallery. On the ground level, there is Bunjilaka, the Aboriginal Cultural Centre, a Children's Gallery, and a Science and Life Gallery. The lower level has an IMAX theatre, and there is a playground outside.

Fig. 3-9 National Gallery of Victoria

South of Carlton Gardens and Melbourne's commercial center as well on St. Kilda Road is the National Gallery of Victoria, the largest museum of art in Australia and the most diverse with respect to its holdings. The location is perhaps even more scenic than that of the Melbourne Museum, since St. Kilda Road is a divided parkway lined with trees and reminiscent of the Paseo del Prado in Madrid. The National Gallery is just on the parkway south of the Yarra River adjacent to the Victorian Arts Centre and Concert Hall. The Gallery dates to 1861 and moved to its present location in 1968. The dramatic new structure of today dates to a 2003 remodeling. Rather more conventional than the Melbourne Museum, the collection in the NGV includes many of the European "old masters," from the 14th through the 20th centuries. There are more than 16,000 such pieces as well as 4,000 Asian pieces. Textiles, furniture, sculpture, and photography are all present. The NGV's Australian art collection is housed up the road and across the river at the Ian Potter Centre in Federation Square.

Fig. 3-10 Ian Potter Centre

The architecture of the National Gallery of Victoria is reminiscent in some ways of the Kimbell Museum in Fort Worth, Texas, in that it gives the impression of solidity and mass. In contrast to the metal and glass of the Melbourne Museum, the NGV is a simple rectangle of stone two stories high. The north side of the structure has a small arched entryway, and the main entrance on St. Kilda Road features a much larger arch under which there is a glass bound wall of water. One enters to the right of the wall and exits on its left. The Gallery is surrounded by a large paved square, which includes a fountain at the main entrance. Above the main walls, there are rhythmically repeated support structures for the overhanging roof. The building itself is, frankly, not especially compelling from an architectural point of view. However, the approach, entryway, and fountain provide just enough elements of beauty to make it an attractive building.

The Ian Potter Centre in Federation Square is a dramatic take on contemporary architecture. Federation Square itself is a complex of buildings whose designs of sandstone, glass, and metal are strongly angular or geometrical. Simple shapes, such as circles and squares, give way to trapezoids and parallelograms. Such a design approach could result in a crazy quilt of shapes, but Federation Square manages to be compelling as a whole. Many of the buildings, including the Ian Potter Centre, also feature panels of mauve, blue, and silver, all of which add to the visual power of the buildings. Even without entering the Federation Square complex, one cannot help being attracted to this quite visually compelling sight.

Where Melbourne seems to be the poster child for dramatic and compelling contemporary museum architecture, Sydney, Australia, has two visually important and persuasive structures in the Art Gallery of New South Wales and the Museum of Contemporary Art. Neither of these structures is contemporary in its architectural style. The Art Gallery of New South Wales is located in the area of Sydney known as The Domain. This park formerly was reserved for the private use of the governor when Sydney was a colony. Today it is a public area adjacent to the much larger Royal Botanic Gardens. This site contributes considerably to the architectural influence of the Art Gallery. Set on a rise in the middle of the park, the Neoclassical building is highlighted for the visitor.

The Art Gallery in its present form dates to 1897. It is an ochre sandstone structure with a central entry portico featuring six fluted Ionic columns. Two more such columns stand at the back sides of the portico, and a shallow but broad stairway leads the visitor to the entrance. There is a typical Greek pediment, but the tympanum is unadorned. The structure emphasizes symmetry and horizontal lines, and at each end of the building are rounded and domed projections with Ionic columns. Although there is less adornment than one finds in a number of other Neoclassical buildings, the Art Gallery is clearly identifiable as a temple of the muses.

Although the Art Gallery of New South Wales houses some European art, the collection is most notable for including the works of European settlers of Australia and those of native aborigines. The former include Tom Roberts's *The Golden Fleece* and Russell Drysdale's *Sofala*, and the latter include an impressive collection of Pukumani Grave Posts done by the Tiwi of Melville Island. There are also extensive collections of photographs, prints, and drawings. The gallery spaces within the building are large and open, making the displays easy to contemplate individually. Across Sydney at Circular Quay is the other architecturally significant art museum, the Museum of Contemporary Art.

Fig. 3-11 Art Gallery of New South Wales

The Museum of Contemporary Art building was built in the 1950s for the Maritime Services Board (http://www.mca.com.au). When the MSB moved to larger quarters, the government gave the building to the University of Sydney, which had a bequest for a contemporary museum from John Power. Power was an Australian expatriate artist who died in 1942. The building was renovated and opened as the Museum of Contemporary Art in 1991. Perhaps because of its original function, the building is not particularly compelling as a museum and certainly not as a temple of the muses. There are, however, a number of architectural features which make the building noteworthy.

First, the location of the MCA is a site of great visual significance. Facing away from the museum, one sees the water of the harbor at Circular Quay and across the way the majestic Sydney Opera House. Second, the museum sits immediately on the harbor, which provides an unobstructed view of the building. Finally, the building is quite large. It is six stories tall and spreads out horizontally so that it has quite strong vertical and horizontal lines. Left and right stairs lead up to the entrance, which is situated centrally at the base of an unadorned central tower. Wings project out from the building on each end. The building, therefore, is quite imposing and thus visually compelling. Like the Tate Modern, the

MCA does not look as if it is a museum, but it has many of the architectural features which make other obvious museums impressive.

Fig. 3-12 Museum of Contemporary Art, Sydney

Like many museums of contemporary art around the world, the MCA has few recognized modern "masterpieces," although its collection does include works by Christo and Jeanne-Claude, David Hockney, Andy Warhol, and Roy Lichtenstein (http://www.mca.com.au). Through the Power bequest, the museum was empowered to purchase contemporary art works and has collected many works by both Australian and International contemporary artists. There are also three collections of varieties of aboriginal art, including a remarkable group of bark paintings. Like most contemporary art museums, the MCA regularly features exhibitions of works on loan. Across the Tasman Sea, art museums with compelling visual power also flourish in New Zealand.

Wellington, the capital of New Zealand, lies at the bottom of the North Island, so it is a seaside city. Many buildings line the principal harbor taking advantage of the scenic location and offering the beauty of the water to the visitor. Many are traditional in their architecture, while some are more contemporary in design. Perhaps the most contemporary such design is that of the Museum of New Zealand, Te Papa Tongarewa. Before discussing its architecture, however, it will be helpful to mention its purposes and contents. This will enable the museum to be correctly classified as an omni-museum and will offer ideas about the relationships between contents and design.

The present museum evolved from earlier, less comprehensive museums beginning in 1865 with the Colonial Museum (www.tepapa, govt.nz/AboutUs). Various museums succeeded this early one until legislation in 1992 established Te Papa Tongarewa. The structure took four years to complete and was opened on Valentine's Day, 1998. The museums which preceded Te Papa had housed a variety of art objects and artifacts, and the new museum amassed them all together under one roof. There are five collections: art, history, Pacific, Maori, and natural environment. Many exhibits are interactive, making Te Papa especially attractive to children. Among these exhibits, for example, is a giant squid. One cannot touch this creature's corpse, but it is possible to stand very close and look directly at the outstretched specimen under glass. Although the body is deteriorating significantly, it remains one of the first things children head for upon arrival.

Fig. 3-13 Te Papa Tongarewa

The art is generally the product either of indigenous peoples or immigrants to New Zealand from Europe. It ranges from the representational all the way to the most abstract. There are a small number of works by the well-known European and American artists, including a Rembrandt etching and a John Singleton Copley painting, but it is not for "old masters" that visitors come to Te Papa. Instead, it is to see the work of Europeans who immigrated to New Zealand and that of the indigenous Maori. If one wishes to study the art of New Zealand on any level, one must visit Te Papa, for it is to be found nowhere else. The collection of Maori art is especially extensive. The culture of the Maori is well-represented by seagoing vessels, huts, and a variety of implements, and there are exhibits showing the history of the islands, immigration, and Pacific cultures generally. Among the items of the natural environment are insects, birds, crustaceans, fish, and so on of the sort one usually finds in a museum of natural history. The collections taken together can truly be described as omnibus.

The builders of Te Papa took great pains to construct a building complex that would be an architectural marvel and one that would in and of itself draw visitors. As noted the complex sits on the city of Wellington's main harbor just southwest of the city center. A small city by comparison to Australian, American, and European ones, Wellington is easily navigated on foot, and Te Papa may be reached from the boardwalk along the harbor or from a city street. Its design is not especially striking from a distance. It is not on a hill, at the end of an esplanade, nor is it a tall structure. While it has five levels, only two are above ground. Once one arrives at the entrance off a large square, the full impact of the building's design becomes apparent however.

The floor plan of Te Papa is unconventional, making it difficult to describe. Essentially it is wedge-shaped, although there are outcroppings on both the street and the harbor sides of the building which complicate the description. No matter, it is the outward appearance of the building which illustrates its architectural impact. As one faces the entrance, there is on the left a two story semicircular wing of beige stone divided in half by a horizontal row of windows. This part of the building at once suggests mass and solidity. On the right of the entryway, there is a narrow wall of rooms of the same height. This wall-like wing is of gray stone and is topped by a roof set at an upward thrust. On the opposite side of this wing are successive large windows along the ground floor. The entrance itself is a wall of windows fronted by an awning of successive v-shaped blue surfaces. The entry hall is spacious, well-lighted by natural illumination, and rises a full two stories. A broad staircase takes the visitor up to the first display galleries. For all its angularity, Te Papa is quite beautiful, and is, in spite of the impressive governmental buildings in Wellington, clearly the architectural jewel of the city.

Christchurch Art Gallery, Te Puna o Waiwhetu, is a more conventional museum than Te Papa, although it houses similar works of art. Located in the largest city on the south island of New Zealand, this gallery in Christchurch is filled with more than 5,000 art objects, including proportionally more European ones than Te Papa. There are no artifacts which would require one to call it an omni-museum however. The main pieces in the collection are by Australian painters. They include Frances Hodgkins's *Pleasure Garden* and *Belgian Refugees* as well as John Reynolds's *Table of Dynasties.* There are a few pieces by lesser European masters, such as Giovanni Piranesi, Antoine Coypel, and Meindert Hobbema.

Fig. 3-14 Christchurch Art Gallery

The Christchurch Art Gallery building is unquestionably the museum's strongest claim to visual impressiveness. It is all curving glass and steel, which is reminiscent of the Melbourne Museum. It is also quite large, spreading out along an entire city block. Along one side, there is a glass enclosed waterfall much smaller than, but nevertheless suggestive of, the larger one at the National Gallery of Victoria in Melbourne. The city of Christchurch is located on the Canterbury Plain, so there are almost no hills on which to erect a temple of the muses, but the Christchurch Art Gallery is set between the city's central business district and the Arts Centre. It has clearly been designed and located to demonstrate the city's commitment to the arts and to stand as a cultural icon for visitors.

Art museums, temples of the muses, are prime examples of the visual persuasiveness of architecture. Regardless of their contents, or whether they be traditional or contemporary, these museums are typically designed to be visually compelling and to complement the aesthetic wonders within. Performance halls, the venues for theatre, music, and dance, are akin to art museums in that they too are temples of the muses. They too are designed to be visually compelling and thus to attract visitors and consumers. Performance halls also must complement the activities which take place within, and their interior spaces must be carefully designed so that the acoustics of the auditoriums and the viewers' lines of sight foster the experience of the performing arts. Performance halls, then, are the next group of structures to be considered as forms of visual persuasion.

CHAPTER FOUR

TEMPLES OF THE PERFORMING ARTS

The art of the theatre is as ancient as civilization itself. Dance, and the evidence for its importance which appears in the art of Greek and Roman culture, may be even older than the theatre. Music, and the lyre and lute of ancient times, has also apparently been with us since humankind began to congregate in groups. Vitruvius (1960) devotes most of the space in Book V of his work to theatres, and amphitheatres are widely distributed across the areas of the Mediterranean which once were part of Greek and Roman civilization. At Lyon in Central France, for example, there are several surviving relics of Roman occupation, including a large amphitheatre. Performances at such places were in the open air, but the rudiments of performance halls were otherwise present: a stage on which the performance was mounted and circular seating in ever-ascending rows to permit all members of the audience a clear view and to insure that even those in the last row of seats would not be out of vocal range. Although amphitheatres were essentially utilitarian, even today's visitors to such places as Lyon will notice the architectural features of curves, arches, and lines used as much to catch the eye as to hold up the structure.

It did not take long for more aesthetic considerations to assume a role in the construction of performance places. The Colosseum in Rome, also known as the Flavian Amphitheatre, was designed and built with an obvious awareness of the visual effect of mass or size. The original structure was perfectly circular, lending symmetry and harmony to its appearance. Its location adjacent to the Forum indicated an appreciation of the importance of site. The repetitive theme of open arches, one row upon another, was a characteristic Roman contribution to the history of architecture. Below the performance level of the Colosseum, there was an intricate group of cells, rooms, and paths to provide for the preparation of performers and mechanisms for delivering people and animals to the surface. While not of visual significance in and of themselves, these places evinced the growing recognition that in temples of the performing arts there would need to be backstage facilities. Historical evidence also shows

Fig. 4-1 Roman Amphitheatre, Lyon

Fig. 4-2 Colosseum

that the Colosseum had canopies to protect audience members from the sun and inclement weather.

These ancient structures hint at the eventual development of what is called here a temple of the performing arts. The evaluation of such temples as means of visual persuasion may proceed largely as it does in other chapters of this book. However, there are features which so distinguish performance halls from almost all other pieces of architecture that special attention must also be given to them. What are they? As implied above, most of these features involve the "backstage." For a theatre production, there must be areas for scene construction, the storage of props and costumes, makeup rooms, and Green Rooms, as they are called by actors and actresses, where performers wait for their entrances. Although the stage itself is primarily an unadorned floor, there must be curtains to conceal backstage work and to provide for entrances and exits, scaffolding, and light fixtures. For such musical performances as those by bands and orchestras, there must be practice rooms and dressing facilities. Opera requires all of the facilities of both theatre and music as does ballet and jazz dance. While acoustics is one of the essential elements of architecture as revealed in Chapter One, it is not particularly important in evaluating architecture as visual persuasion in most of the other chapters of this book. Here, however, acoustics is important, inasmuch as interior appearances of performance are dictated by design elements that will assure good sound transmission and reception.

To this point, the discussion has been on what must go into a temple of the performing arts, but the larger question of visual persuasion pertains more to the audiences which come to performances. The aspects of a performance hall discussed above are actually kept hidden. It is, therefore, the building itself as it is experienced by the audience that should be the focus from here on. Before examining some impressive examples of performance hall architecture, a brief observation about the evolution of performance halls shows again the inclination of humans to design structures not just for their strength and utility but also for their beauty. The early amphitheatres were primarily utilitarian, although there was clear evidence of aesthetic concerns in their designs. Almost immediately, an interest in visual impressiveness seemed to override strength and utility, although not to the detriment of either. A building like the Colosseum was a clear indicator of that trend. By modern times, buildings like the Sydney Opera House make it clear that visual impressiveness or persuasion is the transcendent goal of performance hall architecture.

The Sydney Opera House is so iconic, for that matter, that it is a natural starting point for this assessment. It is a transcendent structure and an icon of contemporary architecture in general. Of course, it is the most recognizable icon of the city of Sydney, and it may be the best known performance hall in the world. The expert evaluation report to the UNESCO World Heritage Committee, which in 2007 named the Opera House a World Heritage Site, said "it stands by itself as one of the indisputable masterpieces of human creativity, not only in the 20th century but in the history of humankind" (http://www.sydneyoperahouse.com). It was designed by the Danish architect and winner of the Pritzker Prize for architecture in 2003 Jørn Utzon and built over the period 1957 to 1973 on Bennelong Point in Sydney harbor in the shadow of the also iconic Harbor Bridge.

Fig. 4-3 Sydney Opera House

The building is easily an example of pure visual persuasion, meaning that it is also a work of art. The clamshell-like roof lines, repeated on the main building and also characteristic of the separate restaurant just to the south, are unlike any other architecture in the world. Their curving lines, ceramic tile covering, and diagonal sweep draw the eye upward. Under the shells, the walls of the building are glass and richly hued dark wood supports may be seen within. As beautiful as the building is, however, its architecture is visually compelling in more traditional ways. First, the site of the Sydney Opera House on a point projecting out into the harbor sets it off from anything nearby. It also makes the building the focal point for ferry riders and those leaving and entering the city via the Harbor Bridge. Second, the building is enormous, something that is not readily apparent in photographs. The width of the base is such that the building has great solidity. Encircling the Opera House is a large open courtyard leading to the Monumental Steps by which visitors ascend to the various venues. The steps are particularly impressive and are 100 meters wide.

The term "opera house" is a misnomer when referring to the building, since it is a complex housing several venues only one of which is an actual opera house. The Concert Hall is the largest of these venues and seats 2679 (http://www.sydneyoperahouse.com). As might be expected from the exterior shape, the Concert Hall has a peaked vaulted ceiling, and the interior is of distinctive brush box and white birch wood. These woods were chosen for their acoustic properties. The Opera Theatre seats 1507. White birch wood is also used here, but the ceilings and walls are painted black. There are smaller venues for more intimate performances. These are the Drama Theatre, the Playhouse, the Studio, and the Utzon Room, each one in this list smaller than the one preceding it. The latter is the only interior space designed by the building's architect. Performances also take place outside on the steps and the forecourt.

The Sydney Opera House illustrates uniquely well the idea of a performance hall designed with the visitor in mind, not only to facilitate an ideal experience of a performance but also to impress the visitor visually. Upon arrival at Bennelong Point, the visitor's first sight of the building, if not breathtaking, is certainly complementary to the idea of fine art being performed within. The building's size reinforces the significance of the goings on. Inside, the rich woods continue the impression of artistry. The woods also enhance the acoustic experience. The exterior is an architectural ornament in itself, and it fairly glows in the bright light of morning and the incandescent light of evening.

Although it is by no means in the same league with the Sydney Opera House, the newly completed James and Nancy Gaertner Performing Arts Center on the campus of Sam Houston State University in Huntsville, Texas, provides an opportunity for a more detailed examination of the elements of acoustics which are important to a contemporary performance hall. It is usually difficult for one to have relatively free access to such buildings as the Sydney Opera House in order to study the interiors and be able to comment on the effect of acoustic considerations on interior design. The new Gaertner Performing Arts Center, however, has just opened at this writing at the author's place of employment, and he was graciously afforded a guided tour of the entire facility by Professor James M. Bankhead, SHSU's Director of the School of Music.

Fig. 4-4 Gaertner Performing Arts Center

Acoustics is of course a science in and of itself and encompasses much information that is beyond the scope of this book. Simply put, however, acoustics in architectural design has to do with understanding and then taking into account the properties of sound as they are affected by the size, shape, and texture of the interior spaces of buildings. With respect to performance halls, acoustic design must also take into account the type of sound most likely to be heard in the building, whether vocal performance or orchestral or small ensemble. Fundamentally, acoustics is concerned with the elements of sound. These include frequency or pitch, the apparent highness or lowness of sound on a musical scale, loudness, and quality. An example from vocal performance will be helpful here.

A singer's pitch is controlled by varying the tension of the singer's vocal folds or cords as they are popularly but erroneously called. Stretching the folds results in higher pitch, while relaxing them under control results in lower pitch. Loudness is controlled by the quantity of air expelled from the lungs. Quality, or the distinctive sound of different voices, is a function of the size, shape, and texture of the resonating cavities through which sound passes. In humans, there are three such cavities: the pharynx, the nasal cavity, and the oral cavity. These cavities shape the outgoing sound by reinforcing some parts of the sound spectrum and damping others. The oral cavity, for example, has both soft (the tongue) and hard surfaces (the teeth) that attenuate some sounds and amplify others. Quality makes it possible for listeners to identify distinctive human voices, in short to recognize a friend on the telephone in the absence of any identifying remarks.

Like human beings, musical instruments are often primarily distinguishable from one another by their different qualities. A trumpet, for example, is sounded by the buzzing of the performer's lips at the mouthpiece, but the sound the listener hears is that buzzing modified by the size, shape, and texture of the air chamber or cone of air between the trumpeter's lips and the bell of the horn. In this way, trumpets do not sound like baritones and basses. Although it is not only the quality that is involved, trumpets also do not sound like oboes or clarinets or pianos. All of this leads one to note at this juncture that an auditorium in a performance hall is a cavity just like the ones discussed above. It, too, shapes the sound of a singer, an orchestra, or a jazz ensemble for the listener's reception. Roth (1993, 103) quotes the architect Louis Kahn as saying that "being in the chamber is like living in the violin. The chamber itself is an instrument." Like a musical instrument, the cavity that is the auditorium must be tuned for the highest fidelity. The challenge for the performance hall architect is to design an auditorium that is optimally configured to reduce defective sound transmission in the first place, to include design elements that will permit the basic structure to be modified or tuned, and to make these acoustic considerations in such a way as not to detract from the aesthetic appearance of the concert hall.

In order to see how the architect accomplishes the tasks just mentioned, it is necessary that the acoustic idea of reverberation be explored briefly. A sound wave oscillates as it travels from the source to the destination. Molecules of air are activated by a vibrating source and radiate out in all directions. Were there no obstructions along the way, a sound wave would eventually decay completely. When a sound wave reaches an obstruction, such as a wall however, it is either absorbed completely, say by a heavy curtain, or it is reflected back along a path determined by the shape of the obstruction. This reflection back is called reverberation. The time required for the reflected sound to die out is called reverberation time. Among other things, it is vitally important that the reflected sound not arrive at the listener's ears at the same time a subsequent sound is radiating outward. Such overlap muddies music and makes speech unintelligible.

"In designing an auditorium, an architect must arrive at the optimum realization of a number of requirements. There should be good sight lines to the performance area or stage; good 'presence,' or strong, even dispersion of initial reflections; good reflection of all frequencies in the sound spectrum; and an even decay of sound during the reverberation time. . . . [and] the optimum time depends on the activity in the room" (Roth 1993, 93). Speech, for example, needs to die out quickly or lengthy reverberation will cause subsequent words to be unintelligible. One second is a desirable reverberation time for speech. Jazz ensembles require about 1.5 seconds, opera about 1.75 seconds, and orchestras some 2 to 2.5 seconds. Before the advent of acoustic measurement, composers and performers alike were forced to adapt to whatever reverberation times they encountered. Large basilicas, with their hard reflective surfaces, were especially hard to adjust to. For many centuries, adaptation was the norm. Today, however, architects working in conjunction with acousticians and using electronic instruments to measure reverberation time are able optimally to design and later to tune contemporary auditoriums.

The Gaertner Performing Arts Center was designed by WHR Architects, JaffeHolden acoustic designers, and Shuler-Shook theatre designers. It is a 100,000 square foot building with a 760 seat Concert Hall, a 175 seat Recital Hall, and a 160 seat Dance Theatre. The first consideration in this assessment will be the interior spaces with an emphasis on the Concert Hall. At first sight, the Concert Hall looks to be a conventional auditorium. There is a stage of blond oak large enough to accommodate a full symphony orchestra. The audience seating is provided by a set of parallel rows rising gently upward. An unelevated mezzanine lies behind this rectangular area. To each side, there are a set of small box seats. A second story of the auditorium contains balcony seating, more boxes, and, behind the stage, a choir loft.

What is unconventional about the Concert Hall, although it is surely to become conventional, are the other parts of the auditorium. These parts are designed to optimize the acoustics of the hall and to permit the chamber to be tuned. The ceiling is open and furnished with the scaffolding needed to mount lights and speakers. It rises another story or so above the balcony and is painted black to minimize the visual effect of the mechanical features above. Just above balcony level at the orchestra seating position, there are curved wooden panels whose positions were carefully adjusted by the acoustic designers to optimize the experience of sound. Set near these panels and at a 90 degree angle to them are other lattice-like panels used for the same purpose.

Fig. 4-5 Gaertner Center Concert Hall

The walls of the hall appear at a distance to curve modestly. On closer inspection, however, it can be seen that they are a continuous series of flat panels set at shallow angles to each other. There are no parallel surfaces anywhere in the auditorium. The optimized reverberation time in the Concert Hall is 2.5 seconds. Along every side of the hall and in the scaffolding above, there are heavy moveable curtains, which can be used to make spot adjustments in reverberation time for performances that have considerable brass and percussion or for performances, such as opera, which combine orchestral and vocal sounds. The acoustical design elements of the Concert Hall are repeated in the two smaller halls.

The moveable panels, grids, and curtains used to provide optimal reverberation time must be carefully integrated by the architect into an interior design that is visually compelling. In short, the architect must take steps to insure that the efforts to optimize the acoustics do not make the hall ugly. In the case of the Gaertner Performing Arts Center, the architects succeeded by using pale blue seat coverings and wall paint, blond wood panels and stage, and by blacking out the machinery directly above the audience. By contrast, some buildings which will be assessed in this chapter have visually stunning ornamental domed ceilings. An example of such a ceiling is to be found in the Royal Albert Hall. However, the beauty of this particular ceiling does not contribute to good acoustics as will be seen. Another example is the ceiling of the Banqueting House in London with a painted ceiling by Peter Paul Rubens, but it too was more decorative than functional from an acoustic point of view.

The interior of the Performing Arts Center is the jewel in its crown and provides the perspective needed in this chapter for assessing those performance halls whose interiors could not be studied so closely. The exterior is nonetheless a visually compelling piece of architecture. Not all of the elements of architecture are at the designer's disposal when preparing drawings for a university campus building. One must usually take the land that is available rather than find what might be the ideal setting. Serendipitously in the case of this center, there was some open land and parking lots adjacent to the university's existing theatre and School of Music, all of which was on an elevated spot overlooking the back of the campus. The site, therefore, is an impressive one. The building itself is a spare, unadorned contemporary design, but it does not fit in with the usual pattern for university architecture. This pattern focuses the limited funds on maximizing the usefulness of the interior space and often results in a simple box.

Fig. 4-6 New Globe Theatre

The Performing Arts Center is a simple, rectangular brick building with a wing at one end full of windows. At the opposite end, a recessed, triangular section provides diagonal lines for the exterior and space for the performing areas within. This section of the structure is colored gray and suggests the idea of soaring to the visitor. Although not part of the architecture itself, there are numerous works of contemporary art displayed within the center, including glass designs, wood and metal creations, and large panels of photographs printed on metal. Since the center is also a working educational complex, there are several open areas for students to congregate. For $38.5 million, Sam Houston State University not only seems to have gotten its money's worth but has vaulted itself into the realm of state of the art for performance halls.

At this point, the essentials of visual persuasiveness to be found in performance halls have been set forth and may be used as a framework for earlier visually compelling temples of the performing arts. Going back quite a bit, Shakespeare's Globe Theatre comes to mind as one of the most unique venues in theatre history. Destroyed by fire in 1613, one today may see a replica of the Globe recently built on the Thames River alongside the Tate Modern at the foot of the Milennium Bridge in London. Frankly, the building is not especially visually compelling when compared to other structures in this book or even in this chapter. It is a simple, circular building much like the Colosseum. It is not ornamented either within or without, and it lacks both color and any rhythmically repeated elemental features.

The Globe, however, does feature architectural elements which set it apart and make it a notable structure. The theatre has an auspicious location, as did the original Globe. The present replica is near the original site on the banks of the Thames (http://www.shakespearesglobe.com). The building is also very large with three levels of seating within, and there is no covering over the stage and yard. In spite of its size, the acoustics within are quite good, probably because of the semicircular seating plan with the stage centered between rows of seating and the open ceiling which eliminates reverberation. The actors and actresses can be easily seen and heard. Indeed, the visually pleasing circular form may have been ideal for presentations prior to modern microphones, speakers, and lighting.

Northwest of London at Oxford University is another visually compelling and essentially circular theatre, the Sheldonian. Adjacent to the Bodleian Library, the Sheldonian Theatre was designed early in his architectural career by Christopher Wren (http://www.ox.ac.uk) following the design of the D-shaped, open air theatre of Marcellus in Rome. A three story Neoclassical structure, the Sheldonian lacks much of the traditional ornamentation of Neoclassical buildings. The lower two stories are defined by rough stone arches which alternate all around with windows and an occasional door. The upper story is a succession of windows separated by pilasters. A balcony rings the roof, and a domed cupola tops the front of the building. The interior has three levels of seating on the floor and two balconies. The second balcony is ornamented with Corinthian columns, and the dome of the structure is painted. While the Globe was not without its creature comforts, the visitor to the Sheldonian could revel in beautiful, comfortable surroundings.

Fig. 4-7 Sheldonian Theatre

Back in London, the best known venue for the performing arts is surely the Royal Albert Hall, which is also circular. While the Sheldonian is a sizeable building set in an auspicious location next to the Bodleian Library, the Royal Albert Hall in Knightsbridge is enormous. It sits across from Kensington Gardens and a monumental memorial to Queen Victoria's consort Albert. The interior auditorium is 185 feet wide and 219 feet long. The exterior was "inspired by the architecture of Northern Italy, was built from some 6 million red bricks and eighty thousand blocks of decorative terracotta. Surmounting the exterior walls and above the balustraded smoking gallery, runs a continuous 800 foot long terracotta frieze composed of allegorical groups of figures engaged in a range of artistic endeavours" (www.royalalberthall.com). The architectural elements of size, line (circle), color, site, and ornament give the Royal Albert Hall its visual impressiveness. Like the Sheldonian, the interior is plush and comfortable. The floor seating is semicircular, and there are two levels of box seating above decorated with Doric columns. Above these levels is a balcony seating level, and there is a rhythmically repeated series of open arches around the top of the auditorium.

In spite of its beauty both outside and in, the Royal Albert Hall was originally quite poor from the standpoint of acoustics (Roth 1993). Roth explains that, although architects have been quite enamored of performance halls with domed ceilings, curved ceilings are usually problematic with respect to sound reverberation, since they do not reflect sound evenly. The problem is exacerbated by a circular floor plan. The Royal Albert Hall had both. Beautiful though both the exterior and interior of the hall were, and beautiful exactly because of their circular shapes, the auditory experience was not harmonious. In order to counteract the uneven sound distribution, a huge drape had to be hung overhead.

Fig. 4-8 Royal Albert Hall

Before leaving London, there is a building which should be examined for its architectural significance, although it is no longer a performance hall. This is the Banqueting House on Whitehall. The structure was built on the site of three other banqueting halls the first of which was part of Henry VIII's Whitehall Palace. At its completion, Henry's palace was the largest in the world. Fire destroyed this complex, so the present Banqueting House is the last remnant. It was designed and built for James I by Inigo Jones, who was a devotee of Andrea Palladio and who had visited Palladian buildings in Italy. Its architectural significance is that it was the first Italian Renaissance structure built in England and designed after the style of Palladio (http://www.timeout.com/london).

As its name suggests, the Banqueting House was certainly used for royal banquets, but it was primarily a performance hall for musical works and especially for masques (http://www.hrp.org.uk). Masques were theatrical performances that usually included singing and dancing as well as acting. They were forms of entertainment for the monarch and his or her court, and indeed the monarch and the courtiers often participated in the masques themselves. The popularity of masques at the Banqueting House, however, quickly fell out of favor when it was discovered that the Rubens paintings which decorated the ceiling of the large double cube in which the masques were performed was being damaged by the soot produced by the candles used for lighting (http://www.hrp.org.uk).

Fig. 4-9 Banqueting House

The interior of the Banqueting House is principally a single, perfect double cube two stories in height. As mentioned, the ceiling is covered by four paintings by Peter Paul Rubens. The coloring of the walls is a pleasant green pastel. Along the first floor wall, there are white fluted Ionic engaged columns repeated rhythmically around the room. Their capitals are trimmed in gold. Two sides of the second floor have a balcony balustrade, and the walls are decorated with Corinthian pilasters accented with gold. Chandeliers now provide lighting. The visual effect is sumptuous and reminiscent of an art gallery featuring old masters. As visually compelling as the interior is, however, it is the exterior of the building which is the architectural jewel.

Inigo Jones's Banqueting House is a two story rectangular building with an undercroft whose façade on Whitehall is as noticeably symmetrical as a structure can be and a fine example of Palladio's emphasis on proportion. The half-story undercroft is faced on the exterior by small square windows and rusticated stonework, giving the base of the building great solidity. Arched and peaked windows alternate along the first story each separated by Ionic engaged columns. The second story has rectangular windows separated by Corinthian engaged columns. The essentially flat roof is ringed by a balustrade. "Jones's Banqueting House is the original result of a calculated combination of Palladio's reconstruction of a Roman basilica after Vitruvius, in plan, and Palladio's two-storey palaces in Vicenza, in elevation" (Cole 2005, 272). Originally, the site was extremely significant, inasmuch as this was the location of Henry VIII's Whitehall palace, but even today the location is auspicious because of its proximity to the Palace of Westminster, 10 Downing Street, and Trafalgar Square. The Italian Renaissance architecture emphasizes horizontal lines, proportion, solidity, and strength without sacrificing delicacy.

Fig. 4-10 Opéra Garnier

As might be expected given the importance of theatre, music, dance, and opera in French history and culture, Paris has two opera houses. These are the Opéra Bastille and the Opéra Garnier, with the latter being by far the most visually compelling piece of architecture as well as the older of the two. Such performance halls are common throughout France as well as Western Europe generally. As will be seen, even a small city such as Nice sports a visually impressive opera house. The 20th century Opéra Bastille is now the home of the principal opera company of Paris, and the older building is now known as the Palais Garnier. It hosts a wide range of performances each year and restorations of both the interior and the exterior in recent years have restored much of the visual power of the elaborate Baroque architecture. Designed by Charles Garnier and built during the reign of Napoléon III at the end of the 19th century, the building is a faithful reflection of the spirit of the Belle Epoque.

Fig. 4-11 Opéra Garnier Interior

Several essential architectural elements come together in the Opéra Garnier. It is not located on the highest point in Paris, but it is nonetheless built on a higher elevation than the city center on the Seine. It also occupies one side of a large square near such shopping attractions as the Galéries Lafayette and commercial banking institutions. However, it is the building itself which would be visually striking no matter where it was located. The building is two stories high but topped with a domed ceiling. The second story is adorned with paired Corinthian columns strung along the façade, while the first story features successive arches reminiscent of the porticoes of Florence.

The interior of the Opéra Garnier is even more lavish than its exterior. The auditorium seats 2,200 and is decorated with red velvet draperies and carpet. The entrance hall is suggestive of the Hall of Mirrors at Versailles in its opulence. There are numerous wall niches holding marble statuary, and the Grand Staircase is dark and polished wood, which begins its ascent upward with a wide expanse splitting into left and right branches halfway to the second level.

Fig. 4-12 Victorian Arts Centre, Melbourne

As mentioned, France has numerous opera houses spread throughout the countryside, including another Garnier designed one in the Mediterranean city of Nice. More than three stories high, the Nice opera house utilizes most of the design features of the building in Paris. It is also on a large square, albeit not at the city's high point. Its most prominent architectural feature is use of Corinthian columns on the second and third stories. The façade is narrower than that of the Paris performance hall, and it features a metal scrollwork canopy over the main entrance. The experience of visually compelling performance hall architecture may be had all over Europe, including Germany and especially Italy. La Scala in Milan, for example, is thought by many to be the quintessential opera house.

It is no less true that far away in the Southern Hemisphere the spirit of visually persuasive architecture may be found in performance halls. The Sydney Opera House is the most striking example of this, but there is a quite striking performance venue in the Federation Square area of Melbourne as well. The Victorian Arts Centre is a complex of buildings, which provides performance halls for symphony concerts, theatre, experimental theatre, and ballet. Its most visually prominent feature is a huge optic fiber tubing spire resembling the Eiffel Tower, which towers over a portion of the complex. The complex itself is largely circular buildings, which are not especially compelling. However, the location of the complex is an architectural gem. Practically on the waterfront of the Yarra River, the buildings adjoin the National Gallery of Victoria, discussed in Chapter Three, and sit on the grand esplanade of St. Kilda Road. On the north side of the river, but within view, are the visually striking buildings of Federation Square.

Fig. 4-13 Lincoln Center

In the United States, the most prominent performance halls are located in Manhattan and in Washington, DC. They are, respectively, the Lincoln Center for the Performing Arts and the John F. Kennedy Center for the Performing Arts. Lincoln Center sits on what was formerly a seventeen city block area of tenements northwest of the city center, which was razed to permit construction of this complex of buildings devoted to the arts and to education. The site alone is therefore quite impressive. The main buildings sit around a huge plaza with a fountain at its center and along Columbus Avenue. Wide stairs lead up from the street to the plaza. Although not designed by only one architect, the structures are certainly complementary in their appearance. Avery Fisher Hall is the home of the New York Philharmonic and is a rectangular structure of concrete and glass some four stories high. It is fronted by a portico defined by a succession of plain columns. Quite similar in appearance is the David H. Koch Theater, which is the home of the New York City Opera and the New York City Ballet.

The Koch Theatre and Avery Fisher Hall face each other across the square, while between them is the Metropolitan Opera House. This building differs from the other two in that its portico is defined by columns that rise to arched dividers. The interiors of these venues are elaborate on the order of such Belle Epoque buildings as the Opéra Garnier. The Metropolitan Opera House, for example, is bedecked in dark wooden trim and red accents. It has a main floor and five tiers of balcony seats. Other buildings in the complex include the Juilliard School, the Vivian Beaumont Theater, and the Time Warner Building. While the United States is a young nation, especially with respect to aesthetic culture, Lincoln Center and the companies which perform there are the equal of any in the world. The architecture is, of course, quite modern, but it is nonetheless as visually compelling as older venues in Europe.

Washington, DC, vies with Manhattan for aesthetic preeminence in the US, and its most serious claim to superiority is the John F. Kennedy Center for the Performing Arts located on the banks of the Potomac River in the Foggy Bottom area of the District adjacent to the now-infamous Watergate. It is a single building, although it provides spaces for numerous, different performances. The companies which perform at the Kennedy Center do not really rival those of Lincoln Center, but there is something always going on from children's programs to jazz festivals to performances of the National Symphony Orchestra and the American Ballet Theatre. The Kennedy Center is the busiest venue in the US.

Fig. 4-14 Kennedy Center

The building is a large rectangle with both riverside and rooftop terraces. It has a flat roof with a smaller rectangular half story centered on it. The ends are without windows, although they are adorned with pilasters. The longer side walls of the building feature vertical windows interspersed with blank wall sections and pilasters. The architecture is distinctly modern and, as such, is clean with strong horizontal lines. The building alone is not especially visually compelling, aside from its size, but its location on the Potomac gives it a site that is quite powerful. Within the Kennedy Center is the huge Grand Foyer the height of the building itself. It is lined with fluted columns and features elaborate chandeliers and a bronze bust of President Kennedy.

As with France, the US is fortunate to have numerous visually compelling performance halls spread throughout the country. Like museums, performance halls are cultural icons such that no aspiring cultural center can long contend for preeminence without one. Aspiring cultural centers, such as Fort Worth and Dallas in the state of Texas, have acknowledged this fact and have moved deliberately toward countering their Western-themed stereotypes with performance halls and complexes that can contend with the very best in other places. In Fort Worth, for example, a centrally located area of the city called Sundance Square has been renovated to attract shoppers, commercial enterprise, government, business, and the performing arts. In subsequent chapters, the courthouse and the skyscrapers in Sundance Square will be assessed as examples of visually persuasive architecture. Here the focus will be on the Nancy Lee and Perry R. Bass Concert Hall.

The Bass Concert Hall, opened in 1998 on the corner of Commerce and Calhoun in Sundance Square, seats 2056 and is able to accommodate many kinds of performances from rock and roll to symphonies to opera and styles as diverse as country and Western music and Classical. Architecturally, Bass Hall is located in a desirable site. Sundance Square is on a hilltop in downtown Fort Worth, and the concert hall, if not in the middle of things, is at least just adjacent to them. Its largely white rectangular shape, three stories high, is adorned with limestone angels 48 feet high on the front of the building. While the remainder of the building is comparatively unadorned, these angels are quite commanding visual presences. They stand on either side of an array of arched windows. There are two entrances to the hall, and between these entrances within is a two story entrance hall decorated with slender columns and an arched ceiling.

Fig. 4-15 Bass Concert Hall

Fort Worth and Dallas are quite close together geographically, but they are not true twin cities. Urban sprawl does mean that one is never outside an urban area while traveling between the two. Like some human twins, Fort Worth and Dallas vie for commercial, political, and cultural supremacy in north central Texas, but each has maintained a distinct character. Fort Worth, for example, is the "Cowtown," known for its stockyards, which even today are a tourist attraction. Dallas, on the other hand, has focused its emphasis on commercial enterprise and has a somewhat more culturally sophisticated image. However, Dallas has until recently lagged a bit behind Fort Worth with respect to the arts. Fort Worth's three prominent art museums—the Amon Carter, the Kimbell, and the Modern—have no real competition in Dallas. To be sure, there is the Dallas Museum of Art, established in 1903, which is now housed in a new modern building. However, the collections in the Dallas museum hardly rival those in any one of the three Fort Worth ones, to say nothing of all three together.

Fig. 4-16 Meyerson Symphony Center

In recent years, Dallas has moved toward a sort of parity with Fort Worth, if not in its museums then in its performance halls. Indeed with respect to its performance halls, Dallas now has surpassed even the formidable Bass Hall through its development of an Arts District for the city adjacent to the downtown commercial center. Home to several venues, the district is most notable with respect to performance halls for the Morton H. Meyerson Symphony Center and the Margot and Bill Winspear Opera House, both of which are beautiful examples of contemporary architectural design.

Little expense was spared in developing the Meyerson Center, which opened in 1989. Its architect was I. M. Pei, the noted designer of the Louvre Pyramids. In many ways the Meyerson Center mimics these pyramids. The hall within is a traditional rectangular design complete with moveable walls for modifying reverberations. It is the home of the Dallas Symphony Orchestra and the Dallas Wind Symphony and is often used by other Dallas musical organizations. The exterior, in keeping both with

Pei's reputation and the notion of modern architecture, is an angular collection of walls and windows integrated with circular, paneled skylights. The effect is inviting and quite visually influential.

Near the Meyerson Center, the Winspear Opera House, opened in 2009, complements the older venue nicely. Designed by Spencer de Grey, the building has a glass façade completely encircling it. Additionally, the Grand Portico, which extends outward from the building on all sides, features louvered coverings acting as solar panels to reduce the temperature surrounding the building. While the principal purpose of the panels and the portico is utilitarian, there is no doubt that it is also visually compelling. The interior is a traditional horseshoe-shaped auditorium. This venue, unlike the Meyerson Center, attracts touring companies of Broadway shows, dance companies, and lectures. It is the home of the Dallas Opera.

Fig. 4-17 Winspear Opera House

Performance halls illustrate in a perhaps unique way the necessity of keeping the use of a building in mind while conceiving its appearance. Notable and visually compelling architecture, such as the Royal Albert Hall, actually sacrificed sound quality to appearance. Particularly in the case of performance halls, it has been crucial to coordinate *venustas* with *utilitas.* If temples of the muses and the performing arts are also archetypal cases of the intent to persuade visually, then surely also are the many houses of worship scattered across the globe. Although this book focuses almost entirely on structures in the Western tradition, it is no less true that shrines, temples, pagodas, mosques, synagogues, and other such places of worship or of religious observance in the East are also intended to be visually persuasive (Ragsdale 2007). To be considered now, then, are houses of worship in Western Europe in the following chapter and others in the English-speaking world in Chapter Six.

CHAPTER FIVE

HOUSES OF WORSHIP: THE CONTINENTAL CHRISTIAN TRADITION

Alongside art museums and performance halls, surely houses of worship rank as preeminent cases of architecture deliberately designed to have visually persuasive effects. Indeed, it is being discovered at this writing that places of worship may have preceded the formation of communities, cities, and governments and might have been the cause of the rise of such things rather than the effect. Conventional wisdom in archeology has held that civilization arose during the Neolithic age of hunter-gatherers and began as communities of people. Once such people had developed cities and governments, then, it has been presumed, they began to build temples or megalithic structures to practice their religious beliefs. With the discovery at Göbekli Tepe, Turkey, of vast temple grounds, the conventional wisdom is being rethought. "The site isn't just old, it redefines old: the temple was built 11,500 years ago—a staggering 7,000 years before the Great Pyramid, and more than 6,000 years before Stonehenge first took shape" (Symmes 2010, 46).

It now appears that the *first* structures were temples or temple-like and that it is those that drew hunter-gatherers together for worship. Places of worship, then, gave rise to communities, to cities, and to governments, not the other way around. Tellingly, these temple-like structures in what is modern Turkey were extensively decorated with bas-relief art, including depictions of cattle, wild boar, leopards, and foxes, as well as snakes, spiders, and scorpions (Symmes 2010). While it is impossible at this point to determine the purpose and significance of these carvings, it is certainly true that houses of worship much more often than not have featured both external and internal art that has a clear theological message.

Of course, Christian houses of worship started as just that: houses. The first Christians gathered in the homes of the followers of Jesus of Nazareth and only much later did Christian communities begin to build what we think of today as churches. By the time of the conversion of Constantine and the exodus of Roman civilization to Constantinople, there began to appear not just churches but large and extravagant basilicas and cathedrals as well as abbeys and monasteries. In virtually every instance, the most visually compelling elements of architecture were incorporated to create a structure worthy, in the builders' minds, of the glory of God. Also in virtually every instance, these houses of worship were ornamented within and without with carvings, stained glass, and sculpture of the highest artistic quality. From the standpoint of the study of visual persuasion, it is especially significant to note that this ornamentation was not merely architecturally compelling but was also theologically significant. The reason is that the ornamentation was very obviously intended to carry a theological message through the depiction of stories from the Holy Bible, of such ideas as the birth of Jesus and the last judgment, and of the lives of the saints.

There are numerous examples of houses of worship as means of visual persuasion in Western civilization. Because of the number of these structures, it seems reasonable to devote a full chapter to the houses of worship to be found in Continental Europe and then to consider in the next chapter those of the English-speaking world, including the United Kingdom, the United States, Australia, and New Zealand. Even so, both of these chapters will merely scratch the surface in their efforts to account for typical examples of church architecture. Since an extensive study of the visual persuasiveness of it has already been made (Ragsdale 2007), the Gothic cathedral of Notre-Dame de Paris is an obvious place to begin the assessment of houses of worship as visual persuasion in the Continental tradition. Subsequently, other houses of worship in France and Belgium, Spain, Italy, Germany, Austria, Hungary, the Czech Republic, and Lithuania will be examined.

French and Belgian Cathedrals and Monasteries

In the previous study of Notre-Dame de Paris as an example of visual persuasion (Ragsdale 2007), the assessment considered the following aspects of the cathedral:

- Windows
- The floor plan
- The geometric precision of the interior and the exterior
- Statuary
- The choir enclosure
- The five portals or entrances and their sculpture

In terms of Chapter One of this book, there are several basic elements of architecture underlying these aspects. The windows, for example, permit a comparatively high level of natural light to illuminate the interior, a specific effect sought by the originator of the Gothic style. Since stained glass was used, the windows allowed the interior to radiate with the colors of the rainbow. Windows and flying buttresses along the sides and back of Notre-Dame provided an obvious visual rhythm. The cruciform floor plan and the geometric precision of the interior and exterior supplied clear lines and symmetry. Sculpture within and without ornamented the structure. Beyond these aspects, Notre-Dame and Gothic cathedrals in general were enormous and usually situated in very central locations.

Fig. 5-1 Notre-Dame de Paris

In the case of Notre-Dame, the location was the Île de la Cité, in the River Seine. In the Place du Parvis, directly in front of Notre-Dame, lies the marker indicating "Kilomètre Zéro," the traditional center of Paris. The location of Notre-Dame was also used for earlier church buildings. The Basilica of St. Étienne had stood there before. So, even the geography contributed to the assertion that the edifice erected in this place was unique. Moreover, as Anderson (1985) notes, "Notre-Dame. . . . was under construction at a period when Paris was the intellectual centre of Europe. . . . [It] is the architectural counterpart of the studium or university of Paris but, as an all-embracing symbol of the age, it encompasses infinitely more than all the treatises on God, Man and Nature written within sight of its flying buttresses and towers" (66).

Several aspects of Notre-Dame's design and construction follow traditional views for the erection of cathedrals. They are 1) the layout of the building in the shape of a cross, 2) the repetition of precise geometric forms in the vaults, windows, and supporting pillars, and 3) the plentiful use of windows themselves. The message of the cruciform layout lies in its imitation of the most familiar symbol of Christendom, the cross of Jesus's crucifixion. The significance of the repetition of precise geometric forms is less obvious. Gothic cathedrals emerged along with the rise of Scholasticism, that blending of Thomas Aquinas's traditional theological views with the rational approach of Aristotle. In Scholasticism, the mind of God could be understood by the rational mind of mankind. Moreover, the mind of God, being itself rational, could be seen in the proportion, symmetry, balance, precision, and beauty of mathematical and especially geometric forms (Scott 2003). All such graphic elements are presumed to be apprehended directly or without central processing.

The architectural details were also supportive of an overall attempt to enhance the light-gathering ability of the structure. Scott (2003) calls light "the most distinctive element of Gothic design" (131). He explains that light was the revelation of God's nature to human beings. If light were the means to apprehend God, then the cathedral conveyed a message to all who entered. The substance of that message was that this place is like Heaven will be. Much as Notre-Dame today seems to the visitor to be imposing, even intimidating, in its grandeur, so much more must it have seemed to the ordinary layperson at the height of *le moyen age*. However, for all of the power of the overarching non-discursive messages of Notre-Dame, it is the specific imagery, or visual sermons, to be found in the statuary and stained glass of the building which must have been of primary importance for such a layperson. Far from mere adornment, the statuary, portal iconography, frieze work, and stained glass were intended as instruction for everyone. This was teaching which could be understood by the illiterate as well as the literate, because it was visual (Anderson 1985).

The stained glass panes of the windows contain a myriad of images from Christian tradition. Images from the Old Testament dominate the rose window in the northern transept, with the center icon being that of Mary with the Christ child. The western rose window has various figures and calendar references. The choir enclosure is a painted and gilded wood frieze depicting, on the north side, twelve events in the lives of Mary and Jesus, for example, the Adoration of the Magi. On the south side, the frieze depicts the nine appearances of Christ following his resurrection, for example, the appearance to Mary Magdalene. In each of these aspects of the cathedral may be found a discursive message aimed directly at the observer. The familiar western face of Notre-Dame illustrates well the idea of "sermons in stone." There are three portals topped by the Gallery of Kings spread symmetrically across the west façade. Each has a tympanum and archivolts filled with sculpture depicting ideas from Christian doctrine. The central portal, for example, depicts the Last Judgment. The left portal depicts the Coronation of the Virgin, while the right is dedicated to St. Anne, the mother of the Virgin.

Not far north of Notre-Dame in a suburb of Paris, there still stands what is probably the very first church erected in the Gothic style: the Abbey Church of St. Denis, commissioned by Abbot Suger, completed in 1144, and now designated a cathedral. The most frequently used term to refer to the great churches of Continental Europe is cathedral, which means that the structure is the home church of a bishop. The term abbey refers to the church of a monastic order presided over by an abbot or abbess. While the terms have ecclesiastical significance, there is no particular architectural distinction. To be precise about St. Denis and many other early Gothic churches, the building was not built from scratch but was remodeled according to Suger's wishes to exhibit the now familiar features of the Gothic style.

Fig. 5-2 St. Denis

St. Denis, remarkably, does not figure in most travel guides nor in much of the literature promoting the French heritage. This is especially puzzling since the church is easily found just off a standard Metro stop. It may be that the working class suburb in which it resides is thought to be dangerous for tourists, but it is quite a contrast to the history of this important structure. St. Denis, whose bones rest here, is a patron saint of France, and for many years the early French kings were buried here. Then, of course, there is the architectural significance of St. Denis.

Although Notre-Dame de Paris is one's usual reference for early French Gothic architecture, one look at St. Denis from any angle reveals the Gothic lines. The western façade, which was the first part of the Carolingian church remodeled by Suger, also has three entry portals although they are not so highly ornamented with sculpture as those at Notre-Dame. Vertical windows rise in two tiers above the left and right portals, while a rose window sits atop the first tier of windows above the central portal. Light and color, the most distinctive features of the Gothic style, are abundant. Unlike Notre-Dame, the church at St. Denis has a tower and spire only over the right side of the west façade. There was originally a matching tower over the left side, but this was removed in the 19th century for a lack of sufficient structural support (http://web.kyoto-inet.or.jp/org). There is also no central spire over the crossing of the nave and the transept.

The removal of the northern tower came too late for there to have been any influence on the proliferation of the Gothic style, but it is unfortunate architecturally. As one looks at the western façade, there is a bothersome lack of symmetry. The upward-reaching verticality is there, and there are elegant flying buttresses to the side, but the abbey church of today is not the complete vision it was meant to be. Like other churches which imitated it's style, however, St. Denis retains the impressive scale and size within and without associated with the great cathedrals and is situated on a higher level than the surrounding buildings.

First, then, there was St. Denis, and then there was Notre-Dame de Paris, but following almost immediately after these two was the church some call "the quintessential Gothic cathedral" (Strickland 2001, 47): Notre-Dame de Chartres. For the visitor coming to Chartres by automobile, the cathedral can be seen from a great distance, because of its situation on what is surely the highest point in the city. It features twin towers on its western aspect topped with spires more like St. Denis than Notre-Dame de Paris. This gives the structure a dramatic vertical line for the eye to follow heavenward. Also like St. Denis, Chartres has no central spire, but like Notre-Dame de Paris it boasts highly ornamented western, northern, and southern portals.

Fig. 5-3 Notre-Dame de Chartres

There are flying buttresses, rose windows on the north, south, and west, and a gallery of kings above the western rose. The northern and southern portals are more deeply set than the western ones and more thoroughly ornamented with sculpture on their tympana and archivolts. Above all, though, Chartres' reputation lies with its stained glass, some 26,000 square feet of it (Strickland 2001, 47). The glass depicts scenes from the life not only of Jesus but of Noah, Joseph, and St. Nicholas. It also depicts everyday craftsmen, such as carpenters and bakers, incorporating the essential quotidian within the spiritual. The only architectural flaw, if it is not too strong a word to use, is that the western façade is asymmetrical. Although there are, to be sure, two towers, the northern one has the appearance of a traditional bell tower and spire, while the southern one is shorter and turret-like.

The growth of the Gothic style was quick and dramatic and, almost coincident with the completion of Notre-Dame de Chartres, Louis IX built a Sainte-Chapelle ("holy chapel") near Notre-Dame de Paris to house what were reputed to be holy relics—the Crown of Thorns and part of the True Cross—which were brought from the Holy Land. Surrounded now by the Palais de Justice, this small chapel is not so distinguished as are other Gothic structures by its size, its site, or by its external ornamentation. Like Chartres, the Sainte-Chapelle is gloriously "rayonnant" because of its many narrow, vertical stained glass windows, which are separated from each other by very fine stonework. These windows have peaked arches and all rise together to a ceiling of colorful blue and gold ribbed vaulting.

Fig. 5-4 Ste. Chapelle

As it is a chapel, not a cathedral, the Sainte-Chapelle is rather small seen from the outside. There is a single entry portal topped with a rose window, no flying buttresses, and no cruciform floor plan, although there is a central spire. The building has two stories, with the chapel itself on the second floor. The architectural effect is one of great delicacy, one of airiness rather than solidity. Visually, its interior may be one of the most compelling in all of Christendom. There are many other houses of worship in Paris and in France that are notable for their use of architecture as a means of visual persuasion, but the final two for the purposes of this account are Le Mont St. Michel, a Romanesque monastery on France's northwest coast, and the great coronation church of French kings: Notre-Dame de Reims in the northeast.

Le Mont St. Michel is surely familiar even to those who have never seen it, because of its popularity as a picture postcard subject and one of France's most frequently visited destinations. Like Chartres, le Mont St. Michel can be seen from afar sitting atop its "mount" just off the coast of Normandy and west of the World War II invasion beaches. These beaches are also those from which Guillaume le Conquerant sailed to England in 1066 CE. Although the monastery is its *raison d'etre,* the entire complex is actually a fortified city complete with castle walls. Although there is now a road from the mainland, le Mont St. Michel is on an island completely surrounded by water at high tide. In all the history of France, this fortified emplacement has never been breached by an enemy.

Fig. 5-5 Le Mont St. Michel

The most significant architectural feature of Le Mont St. Michel is its site. While the monastery and the walled city surrounding it are compelling, they would undoubtedly not be so were they not located on a mountain on an island. Additionally, the structures fully cover the island, making it seem as if they have been thrust up from the sea itself. There is no nearby town to obscure the view, so as one approaches by auto from the east there are vast open fields followed by open water. Only sheep grazing draw attention from the mount. The visitor passes through a portal in the fortifications and then proceeds to make a rather sharp ascent to the monastery, now passing numerous shops, hotels, and restaurants.

Arriving at the west entrance of the monastery itself, the visitor encounters a rather plain single portal flanked by arched bays with windows. Across the front are Corinthian columns, but there is no sculpture. Rising impressively above the crossing of the nave and transept, there is a massive tower and spire, which enable one to distinguish the monastery from afar from the surrounding buildings. Closely spaced flying buttresses support the area of the choir behind the spire. This area, or chancel, was added to the monastery in the 15th century and was done in the Gothic style. The buttresses are adorned with spires decorated with sculpted flowers.

Within the church, columns and walls are of unadorned stone, giving the place of worship an especially austere look rather typical of the interior of a Romanesque church. The nave rises to ribbed vaulting above. The chancel or choir, following the later Gothic style, is brilliantly lit by three stories of narrow vertical windows separated only by thin stone frames. On the north side of the church is the cloister, which is perhaps more compelling architecturally than the church itself. This courtyard is surrounded by a covered walkway with double columns. Each of these thin columns connects to the others by graceful pointed arches. The spaces between the arches are adorned with rose or leaf motifs. The beauty of this place and the impression it leaves mark Le Mont St. Michel as one of the most compelling of all the architectural sights of Europe. Although much more like the cathedrals of Paris and Chartres than the monastery of Le Mont St. Michel, Notre-Dame de Reims is impressive in its own right.

Fig. 5-6 Notre-Dame de Reims

Even were it not the cathedral where most of the kings of France were crowned, Reims would stand out for its ornament and its faithful rendition of the Gothic style. In roughly the city center of Reims and at the top of a gentle rise, the cathedral is approached by a cobblestone street. The western façade closely resembles Notre-Dame de Paris with three portals, a large central rose window, a gallery of kings, and two towers. There is a spire over the nave and transept crossing, but it is not so high as the one in Paris. Although the archivolts of the three portals are ornately sculpted, along with similar jamb statues, the tympana are windows rather than friezes. The towers have especially delicate and open stone tracery. The flying buttresses around the choir, like those at Le Mont St. Michel, have spires atop them. The interior of the Reims cathedral soars to ribbed vaulting above narrow columns and is colorfully lighted by stained glass in the windows of the choir and in four rose windows: two above the central west portal and one each over the north and south portals. This account of compelling church architecture could continue substantially, inasmuch as there are so many cathedrals to visit in France. Before leaving this area of Europe, and because of architectural similarities, Brussels, Belgium, should be the next stopping place. The Cathedral of St. Michael and St. Gudula is another compelling example of the Gothic architectural style.

The architectural power to influence often begins with an elevated location, and St. Michael's is no exception. Brussels is divided into the Lower Town, where one finds the Grand Place with its City Hall, and the Upper Town of the Royal Palace and the art museums. St. Michael's is located in the Upper Town and commands a clear overview of the older part of the city below, including an unobstructed view of the spire of the City Hall. The cathedral is done in the Gothic manner, although its interior has features that are normally thought of as Romanesque. The western entry façade has a large central portal with double doors and two smaller flanking portals. Unlike the deeply recessed portals of many Gothic cathedrals, these three are shallow. Each rises to

Fig. 5-7 Cathedral of St. Michael and St. Gudula

a peaked arch, and each has some simple sculptured adornments. There is no rose window, but the central portal is graced with tall double windows with peaked arches.

Rising four stories in the air, two square bell towers complete the western view. The western entryway is symmetrical and is fronted by a small park leading away toward the Lower Town. Unlike a great many old stone buildings in Europe, the stonework of St. Michael's is clean and attractively white in its appearance. The interior of the cathedral is highly illuminated in the Gothic fashion, but the support columns along the nave are unadorned stone of sizeable proportions much like those in Romanesque buildings. They are separated by peaked arches and above them the ceiling is ribbed vaulting. While the columns themselves are bare stone, they do have individual sculptures attached, which face the nave. Having begun this account of houses of worship in the Continental tradition in France and Belgium, it is now time to examine a similar sample of the houses of worship in Italy.

Italian Cathedrals

Italy, like France, is home to some of the most famous houses of worship in the world, although as in the case of the city of Pisa, this fame is sometimes less the result of compelling architecture than it is the outcome of construction gone wrong. Indeed, Pisa is the place to begin in an effort to see the Italian contribution to religious architecture and visual influence. One virtually unique feature of Italian church architecture to be seen in both Pisa and Florence is the three structure complex where separate bell towers and baptisteries sit alongside the cathedral itself. In some cases, the cathedral is the focal point, while in others it is the bell tower. In Pisa, for example, it is the bell tower, which began to lean noticeably shortly after its construction, that is sought after by tourists. It is not just this structural anomaly which makes the tower compelling though, since it may also be the most delicately beautiful such structure in all of Italy. In Florence, the Basilica di Santa Maria del Fiore, or Duomo, is striking for its dome, its baptistery for its carved bronze doors, and its campanile or bell tower for its design by Giotto, so that there is no one focal point.

Fig. 5-8 Cathedral and Leaning Tower of Pisa

Begun in the mid-11th century, the cathedral of Pisa is a Romanesque building but one which bears little resemblance to those in either France or England. There is a cruciform floor plan along with colonnaded aisles and numerous small clerestory windows. Unlike many Romanesque churches, the columns are slender, ornate, and connected by curved arches, which perhaps prefigure the similar style of Florentine Renaissance architecture. Moreover, the site—the Campo dei Miracoli—is rather higher than the surrounding parts of the city and commands an unobstructed view. The cathedral's beauty and the real source of its visual power is the entry façade. Across its face, three portals separated by Corinthian columns stretch rhythmically. The portals are rather plain and without sculpture, but above them, stacked one upon the other in successively decreasing width, are four stories of slender columns much like those in the interior. Although without the sculpture of such cathedrals as those at Paris, Chartres, and Reims, the façade is richly decorated with marble and mosaics with glazed inlays.

Begun a century after the cathedral, the round bell tower is decorated in harmony with the entry façade of the church. Circling each of the tower's seven stories—the bell chamber is an eighth and smaller story—are the same slender columns separated here by curved arches. In spite of its having become a curiosity because of its lean and a historical site because of Galileo's gravity experiment, the Leaning Tower of Pisa is a compelling visual form. It is massive without being overwhelming and at the same time delicate in its ornamentation and rhythm. Although much less interesting architecturally, the baptistery is a free-standing dome surrounded by the slender columns of the cathedral and bell tower. Were it not for the other two buildings, this one would certainly be more prominently discussed. Its uppermost story, just below its dome, has pointed arches as one finds in the Gothic church style.

Fig. 5-9 Duomo

Southeast of Pisa in the cradle of the Renaissance is the city of Florence, the center of which remains today much as it was in Dante's time. Undoubtedly the city's most striking architectural feature is the Basilica di Santa Maria del Fiore, more often called simply the Duomo with its dominating central dome designed by Brunelleschi. Although not located on a hill, the site of the Duomo, its bell tower designed by Giotto, and baptistery with its bronze doors by Ghiberti, is in the city's center. There are no other buildings to obscure the view, therefore the Duomo may easily be seen from great distances.

Built in much the same architectural style as Pisa's cathedral, the Duomo has a rather flat entry façade. There are three portals topped with pointed arches, the north and south ones of which are topped with small rose windows. The look changed considerably in the 19th century, when the façade was covered by panels of red, green, and white marble giving the Duomo the appearance of Giotto's campanile. While the façade is now quite colorful, a desirable feature architecturally, it borders on the gaudy and is less compelling than it might have been. The dome, of course, is both tall and vast, covering an equally spacious and empty interior. Within, the ceiling of the dome is covered with frescoes of the Last Judgment begun by Vasari and completed by Zaccari. Both the exterior and interior size of the dome are its compelling architectural features, so that one hardly notices the perhaps too decorative façade.

Fig. 5-10 Giotto's Campanile

The campanile is a square tower rising above the height of the church but not above the dome. It, too, is highly ornamented but lacks the graceful columns of the Leaning Tower in Pisa. It conveys a sure sense of mass and weight. The baptistery is an octagonal structure whose claim to visual influence is its eastern doors. Done in bronze bas-relief by Ghiberti, ten panels depict some of the principal stories of the Old Testament. To preserve the originals, copies now grace the doors, and the originals are held in the Duomo's museum.

To the northeast of Florence lies the former city-state of Venice, famous for its waterways, vaporettos, bridges, and above all for the Basilica of San Marco, its piazza opening onto the Venetian lagoon, and the adjacent Palazzo Ducale or Doge's Palace. For many years, Venice was one of the leading maritime powers of Europe with trade extending both east and west. Because of its commerce with the east, influences of Constantinople and Byzantine architecture are clearly evident in Venice's basilica.

Fig. 5-11 Basilica of San Marco

The entry façade of San Marco is expansive, with five entry portals. Each portal includes a rounded arch and is decorated with two levels of slender columns. The arches themselves each contain colorful mosaics, and the main portal is superposed by four bronze horses brought to Venice from Constantinople, although the present ones are copies of the originals. The second story above the entry portals repeats the rounded arch pattern and the colorful mosaics. Between each second story arch rises a pointed spire, and these spires continue along the sides of the church as well. The floor plan is that of a Greek cross topped with a large dome over the crossing of the nave and transept and four smaller domes over the apses.

In typical Byzantine fashion, the interior is brilliantly decorated with colorful mosaics in the domes depicting Pentecost and the Ascension among other biblical stories and with gold paneled altarpieces. Strickland (2001, 36) describes the interior this way: "Glowing color and gold-sheathed vaulting dissolve surfaces to create an earthly vision of heaven. Forty thousand square feet of gold-inlaid mosaics and 500 plundered pillars of jasper, porphyry, and marble adorn the interior. Above the altar is an enamel and gold screen encrusted with hundreds of pearls, garnets, amethysts, emeralds, and rubies."

The principal compelling architectural features of San Marco are surely the size of the building and its rich ornamentation. The entry façade also stretches from side to side in symmetrical fashion and with the rhythmic repetition of its slender columns. The present campanile dates from the 20th century, replacing one that had fallen. There is no baptistery. For its Byzantine architecture, San Marco is virtually unique in western Christendom and in Italy. No cathedral, however, in Italy or elsewhere is so architecturally compelling as the greatest Christian house of worship in the world: the Cathedral of St. Peter in Vatican City.

Built over the tomb of St. Peter and in the location of the Circus of Nero, today's Cathedral of St. Peter, the church of the Pope, is an Italian Renaissance structure with Baroque features. Its dome was designed after the Pantheon by Bramante but redesigned later by Michelangelo. The floor plan is basically a Greek cross with an extended façade and narthex. Within, the canopy above the tomb of Peter was designed by Bernini. It is not possible or even desirable to identify which of St. Peter's many architecturally compelling features is the most impressive, but the very first impression by the viewer is probably its size.

The approach to the entryway of the cathedral is through a huge circular piazza designed by Bernini. Semicircular walls surround the piazza consisting of colonnades topped with statuary. In the center of the piazza stands an obelisk removed from Egypt and two fountains. The semicircular colonnade opens into a rectangular area leading to the church's entrance. That entrance, designed by Maderno, is fronted by broad stairs and spreads from side to side symmetrically with three entry portals separated by Tuscan columns. Above is a second colonnaded story topped with statuary. The central dome itself, like Florence's Duomo, dominates everything, with paired columns at its base and a colonnaded turret at its peak.

Fig. 5-12 St. Peter's Basilica

The design elements which make St. Peter's so compelling may be seen in other familiar structures both ecclesiastical and secular. St. Paul's in London, the American Capitol in Washington, DC, and the Texas State Capitol in Austin all resemble St. Peter's. While the decoration of the interior is not so rich as that at St. Mark's, there is nothing to match Bernini's canopy, nor is there a Pietá to match the one by Michelangelo. Additionally, if the exterior is impressive for its mass, then the interior is equally so for its spaciousness. It is the largest such interior in Christendom, and there are marks on the floor showing where other famous houses of worship would fit. Before leaving southern Europe, there is another significant, visually compelling cathedral to examine in Spain.

Spain

The Spanish capital in Madrid is home to several notable examples of visually persuasive architecture, including The Prado, the Palacio Real, and the Cathedral of Santa Maria la Real de la Almudena. The latter is a comparatively recent addition to the European landscape. It was begun in the late 19th century, but its construction was stopped by the Spanish Civil War in the 1930s (http://www.sacred-destinations.com/spain). Work began anew in 1944, and the cathedral was completed in 1993. The architecture is basically Neoclassical with Baroque touches, but the interior mimics Gothic cathedrals. Sitting directly across a huge plaza from the Palacio Real, the cathedral's most notable feature may well be its size. Both it and the Palacio sit atop a tall outcropping of rock on the western edge of Madrid, and the site is therefore particularly impressive.

Fig. 5-13 Santa Maria la Real de la Almudena

The broad horizontal line of the Santa Maria cathedral imparts a visual solidity. In the center of the symmetrical façade, the central entry portion of the structure rises some four stories and is flanked with two bell towers. Behind and between these towers is the cathedral's central dome, and between each tower are two story colonnades. The lower one is of the Doric order, while the one above is of the Ionic. The architectural elements which make Santa Maria la Real de la Almudena so visually compelling are, again, site, size, symmetry, and ornamentation. The structure works quite well as a visual complement to the Palacio Real across the plaza. From the approach to both structures from the southeast, the cathedral can be easily seen from a distance.

Eastern and Northern European Cathedrals

Surely, France and Italy are home to the most visually compelling houses of worship in all of Continental Europe. However, it would be misleading to fail to notice several other quite visually distinctive houses of worship at Prague and Kutná Hora in the Czech Republic, at Esztergom and Budapest in Hungary, and at Vilnius in Lithuania. Most of these houses of worship have been obscured through the political and geographical isolation of their countries following World War II, but they are readily accessible today. The Cathedral of St. Vitus in Prague is virtually the symbol of the city and a standard postcard subject, inasmuch as it stands atop the highest point in the city and is surrounded by Prague Castle and its fortress walls.

Fig. 5-14 Cathedral of St. Vitus

St. Vitus has a majestic appearance, partly the consequence of its site but probably more because of the cathedral's spires. Twin Gothic spires rise above the western portals, and a central spire of Baroque design is attached to the southern side of the structure. The floor plan of St. Vitus is not cruciform but rather is a simple basilica. Pinnacled flying buttresses adorn the choir as at le Mont St. Michel. The three western portals have peaked arches and are recessed enough that their archivolts may be adorned with sculpture. Above the central portal is an enormous rose window. Within the cathedral, the nave is lit by stained glass windows in the choir and along the sides. The colonnades are simple but not massive.

Fig. 5-15 Cathedral of St. Barbara

A short distance from Prague, in the small city of Kutná Hora, is the architecturally unusual, but basically Gothic, Cathedral of St. Barbara. Not located on a hill, rather in a small valley, St. Barbara is remarkable for two features: the many flying buttresses surrounding both the nave and the choir and three central spires. Viewed from above, the cathedral has been likened to a royal crown, and the pointed spires associated with the buttresses certainly suggest the many points of a crown. Unlike other Gothic cathedrals, St. Barbara's western façade has no towers, no spires, no rose window, and only one central portal largely unadorned.

In the neighboring, former Soviet bloc, country of Hungary, there are several striking examples of visually compelling architecture to be seen in the capital of Budapest. However, it is probably at Esztergom, north of Budapest, that one will find the most striking cathedral in the country. The small town of Esztergom sits on the Danube River bend border with Slovakia. High on Castle Hill above the Danube is the 19th century Neoclassical Esztergom Cathedral. The cathedral's full name is the Primatial Basilica of the Blessed Virgin Mary Assumed into Heaven and St. Adalbert.

The approach to the cathedral is a wide paved walk leading upward. The Neoclassical façade has eight Corinthian columns topped with a traditional Greek pediment. Rising above the center of the building is a dome ringed with smaller Corinthian columns much in the manner of London's St. Paul's. To the left and right of the central dome are two smaller wing domes, which accentuate the structure's symmetry. Within, the cathedral is richly adorned with marble columns topped with Corinthian capitols in gold. The arched ceiling and the wall behind the choir feature paintings of religious subjects.

Fig. 5-16 Esztergom Cathedral

In the capital city of Hungary, Budapest, stands Matthias Church, technically the Coronation Church of Our Lady Wicha, high atop the hillside part of the city which was once the separate town of Buda. Not only is this an architecturally impressive location, but it also overlooks the magnificent Danube River. Although surrounded by many other structures in this old part of the city, Matthias Church is set off by an open square in front and spacious areas to the sides. Its most noticeable feature, although surely not its most impressive, is a gaudy diamond patterned tiled roof. The church is essentially a Romanesque basilica with a Latin cross floor plan. However, there are three chapels on each side of the building which extend out from the main body of the structure. Above the crossing of the nave and the transept, there is a small spire.

The entry façade of the church is its most striking architectural feature. Resembling Notre-Dame de Chartres, there is one richly adorned Gothic spire rising vertically toward Heaven on the right as one faces the entrance and a much smaller circular spire on the left. The large spire is richly decorated in the Gothic manner with multiple statues, while the entrance itself is surrounded with delicate stone tracery that has the appearance of filigree. Above the entrance is a small rose window.

Within Matthias Church is a highly decorated interior in rather a Byzantine fashion. The high altar is richly illuminated by two stories of stained glass, and the walls of the church are richly painted with religious scenes of varied colors but with a primary emphasis on gold. The location of Matthias Church is replete with many other architectural attractions, as will be seen subsequently, but the church itself is easily its most compelling example of architecture.

Fig. 5-17 Matthias Church

Vilnius, the capital of Lithuania, is well off the beaten tourist path and a city little known and perhaps little regarded in the West. That is a sad commentary, since this lovely city is home to a magnificent university, an imposing castle, and many, many churches and cathedrals. Two of the most visually interesting of the latter are the Vilnius Cathedral, an imposing Neoclassical structure in the center of the old city, and the Cathedral of St. Peter and St. Paul, a richly adorned Baroque structure southeast of the city center.

Vilnius Cathedral draws much of its architectural power from its centrally located site. It is also surrounded by open spaces, so that the effect of the building on the viewer is enhanced. The entry façade is strictly Neoclassical, with a portico supported by six Doric columns and a Greek pediment. Similar but smaller columns continue along each of the sides of the cathedral. Set back from the portico is the building itself with a single simple doorway but with rounded alcoves to each side filled with statuary. The building is a simple rectangle, but chapels and a sacristy extend out at both ends of the rectangle to give the floor plan the shape of the capital letter "I." The columns and pediment are of off-white stone, while the building itself is of a yellow beige stone. On the right side of the western façade, there is a four story belfry of white and beige stone.

The interior of Vilnius Cathedral is a simple basilica with the usual columns along the nave. The exterior color scheme continues inside, with the cream colored walls and columns. The columns are adorned with hung paintings, and the arched vaults above the columns are covered with successions of carved medallions. At the altar end of the nave, there is a columned entryway topped with an ornate silver and gold organ face. The chapels are filled with statuary, marble walls, and statuary, some of the latter of which is gold. The interiors of these chapels are Rococo rather than Neoclassical.

Fig. 5-18 Vilnius Cathedral

Vilnius is a city of churches and cathedrals. Although it is not true, as is popularly remarked, that there is a church on every corner, there are certainly a plethora of them. Of these, the one other to be noted here is the Baroque Cathedral of St. Peter and St. Paul, on a gentle rise near the River Neris. Although not situated so prominently as it might be for architectural impact, this cathedral nonetheless has an imposing presence. It lies behind a stone and metal fence around the perimeter of its grounds, which isolates the building from the surrounding structures. Unusual for a cathedral site, there are trees within the walls and in front of the building, so that the view from the entry façade appears almost to be a residence rather than a house of worship. However, the impact on the viewer is not entirely diminished, and the citizens of Vilnius regard the cathedral as "the most exquisite Baroque monument in Vilnius" (Venclova 2002, 175).

Fig. 5-19 Cathedral of St. Peter and St. Paul

The entry façade is two stories high topped by a cupola and flanked by two small towers. Both the entry portal and the story above are adorned with small Ionic columns. The stonework, like that at Vilnius Cathedral, is largely beige. The floor plan, also like Vilnius Cathedral is a simple rectangle with chapels and sacristies front and back. The nave is flanked by high arched passages into various altars on each side, and the high altar sits in a domed alcove. The adornment is typically rich Baroque. The walls, passageways, and domes are filled with statues, carvings, and gold cladding. Perhaps the most striking adornment is a large gold mesh chandelier in the shape of a ship. While this Baroque vision of what Heaven must be like differs dramatically from that portrayed in the Gothic style, it is no less impressive. This cathedral does not draw the viewer's eye upward toward God so much as it provides a feast of color and ornament. As such, the Cathedral of St. Peter and St. Paul in Vilnius is a good introduction to the newly restored Frauenkirche in Dresden. Before examining that house of worship, however, it is Vienna to which one should turn for another taste of the Gothic style.

Austrian and German Cathedrals

Austria and Germany are tied together not only by their common language and cultural heritage but also by their similar religious traditions. Germany, the home of Martin Luther, is the birthplace of the Protestant Reformation, and there are prominent Protestant houses of worship to be found throughout the country. In both Germany and Austria, there are numerous examples, as there are in France and Italy, of the Gothic style of cathedral building, many done in the years before the Reformation. The Cologne cathedral is one such house of worship that immediately comes to mind. However, two of the most architecturally interesting and visually compelling of the Austrian and German cathedrals are St. Stephen's in Vienna and Dresden's newly restored Frauenkirche.

Like most capital cities, Vienna, Austria, features a variety of architecturally impressive buildings. Surely, however, it is the Cathedral of St. Stephen, with its soaring Gothic spire on the south side that is the epitome of compelling architectural form in the city. A much shorter spire stands on the north side. The south spire is richly ornamented, much as the spire is at Matthias Church in Budapest. The arched west entryway to the cathedral is less Gothic than Romanesque, with twin spires on either side. Again like Matthias Church, the triangular pitched roof is covered with colorful glazed tiles set in place to represent the Habsburg double-headed eagle and a general diamond pattern. The church, of golden limestone, has for many years carried the black soot from the chimneys of the city, but it has recently been cleaned without and within.

Fig. 5-20 St. Stephen's

Not only is the south spire vertically eye-catching and the roof richly ornamented, St. Stephen's is a massive church. Today, it is crowded by a clutch of surrounding buildings, but its size is an essential architectural feature contributing to its compelling form. Inside, the Romanesque nave has richly decorated, almost Baroque, walls and a diamond patterned floor. There are 18 altars, with more in chapels along the sides, and the pulpit is of intricately carved stone in the Gothic style. The figures which adorn it are St. Augustine, St. Jerome, and St. Gregory. St. Stephen's, in short, is one of the most visually compelling houses of worship in all of Christendom and famous enough as a landmark to appear on the Austrian ten cent euro coin.

North of Vienna, in the city of Dresden, Germany, is the final house of worship to be considered here. Dresden, of course, is most famous for the tragic fire-bombing to which it was subjected near the end of World War II, although before that it was widely regarded as one of the great cultural centers of Europe, a Florence on the Elbe. While there has now been great progress made in repairing the extensive damage from the bombing, there is no greater evidence of the transformation than the rebuilding of the Lutheran Frauenkirche, or Church of Our Lady, which had been leveled by the firebombing. Not hit directly by them, the firestorm which accompanied the bombs caused temperatures within the structure to rise so high that it imploded. Since Dresden lay within the boundaries of East Germany during the Cold War, there was no special priority given to resurrecting a church, although to be sure the Soviets did repair damage to the palace, the opera house, and the museums of the city. Since 2005, however, the Frauenkirche has been open in all its restored splendor, and splendid it certainly is both within and without.

Fig. 5-21 Dresden Frauenkirche

The Frauenkirche is located in the northern part of Dresden's Old Town near the Elbe River and situated in a large open square between the Royal Palace and the Albertinum. The church was and is a sight to see in architectural terms, for it is essentially that most structurally sound of building elements: a square. The design is Baroque, and at each corner of the square there are small towers. The overarching architectural feature is the huge central dome which covers the entire building space below. It is topped by an open cupola, giving the entire structure the appearance of a huge handbell. Original sandstone pieces left as rubble were carefully preserved over the years of the Soviet era and eventually incorporated into the reconstructed church. The new construction is of beige sandstone, but the original pieces were left uncleaned in order to be able to identify them on sight. From any direction of approach, the Frauenkirche has solidity, if not great mass, it is bright and clean, and it draws the viewer's eye upward.

The Baroque interior, a single domed room as in Rome's Pantheon, is a symphony of rich, dark woods and white walls and columns ornamented with statuary and gold decoration. The area behind the altar is particularly ornate with fluted Corinthian columns and gold capitals, statuary in stone and gold, and organ pipes crowning everything. Within the sanctuary are three levels of balconies with brightly painted fronts. Equally brightly painted columns support these balconies. The inside of the dome is painted with eight scenes from Bible stories. The Baroque style is regarded by some as gaudy and perhaps tasteless. Taken as an example of the builders of houses of worship to represent the Heavenly Jerusalem, however, it is probably better seen as a treat for the eyes.

Continental Europe is a treasure trove of houses of worship whose architecture is compelling. The visual persuasiveness of these buildings is rooted in such fundamental architectural elements as line, rhythm, scale, space, mass, and, most importantly, site, light, color, and ornament. The churches and cathedrals of the Continent continue to hold forth the belief of their builders that the house of God should symbolize the nature of Heaven, should be ornamented in glass and stone with representations of biblical messages, and should stand alone as a mute example of God's power and majesty. This summary is no less true of the houses of worship to be found in the United Kingdom, the United States of America, and in Australia and New Zealand, and it is to those structures in the English-speaking world that attention is now turned.

CHAPTER SIX

HOUSES OF WORSHIP: THE ENGLISH-SPEAKING WORLD

When one thinks of the great churches in the English-speaking world, surely it is London that first comes to mind with its world renowned Westminster Abbey and St. Paul's Cathedral. Of course, these two structures are only the most famous of a large array of houses of worship in the capital city, many of which were designed by Sir Christopher Wren. It would not be difficult to devote an entire chapter of this book to the churches and cathedrals of London. Instead, however, as with Continental Europe, attention will be given to houses of worship throughout the United Kingdom, which are dramatic examples of the power of architectural influence. After London, churches in Bath, Salisbury, York, Durham, and Edinburgh will be examined before crossing to the US and ultimately traveling down to Australia and New Zealand. In a subsequent chapter dealing with universities, attention will also be given to chapels incorporated within the confines of large universities.

British and Scottish Cathedrals

Although the Protestant Reformation originated with Martin Luther and the 95 Theses he posted on the doors of the Schlosskirche in Wittenberg in 1517, European houses of worship were largely untouched physically by the schism. Protestants simply went about their business by building their own churches. In England, however, at least two events led directly to the physical destruction of church property, including buildings, statuary, stained glass, books, altarpieces, vestments, and other trappings of religious orders. These were the Suppression of the Monasteries by Henry VIII and the iconoclastic movement in the following century (http://www.luminarium.org/encyclopedia/dissolution.htm).

Luther's ideas were known and discussed in England, but it was probably the headstrong assumption of the headship of the church by Henry VIII which set the country on a path leading to a split with Rome and the establishment of Anglicanism. During the period 1536-1541, Henry actively suppressed the English monasteries, confiscating their buildings, land, and possessions and dissolving their orders. There were many motives for these actions, including diverting wealth and power to the king and diminishing the influence of the Church.

Fig. 6-1 Westminster Abbey

During the following century, civil war brought further destruction to cathedrals, abbeys, and monasteries. Soldiers took them over to serve as fortresses, and other soldiers brought down their walls. Moreover, an iconoclastic movement among the burgeoning ranks of Protestants brought the scourging of emblems of Catholicism such as statues, altarpieces, stained glass, and the like. These historical events are especially significant in a contemporary assessment of cathedral and church architecture as visual persuasion, since so many of the great English cathedrals, originally Catholic, stand today stripped of some of their most important adornments. Empty alcoves and pedestals are all that remain of the original installations, a fact that is strikingly noticeable in York.

Westminster Abbey, actually the Collegiate Church of St. Peter at Westminster, is the burial place of kings, queens, and other notables and the place of coronation of English kings and queens. It stands in the shadow of the Palaces of Westminster, the seat of British government. However, the architecture of Westminster Abbey does not take second place to its historical and political significance. It is a magnificent structure, not only for its location but for its primarily Gothic style. The building is laid out in classic Latin cross fashion, with a westward facing entryway. There is a single portal flanked by two towers. There is a row of statuary above the door surmounted by a tall, arched group of windows. Since the visitor today typically enters the church through the north door, it is this entrance which is the more familiar one.

Fig. 6-2 St. Paul's Cathedral

The north entrance is most evocative of the western façade of such cathedrals as Notre-Dame de Paris and Notre-Dame de Chartres. It is symmetrical, with three entry portals and doors deeply recessed within peaked arches. The tympanum of the central portal is richly adorned with statuary, but alcoves along the ground level are empty of statues. Four pilasters rise skyward between and flanking the portals, and these are adorned with statues in the upper levels. Atop the central portal is a large rose window.

The nave rises some ten stories, certainly suggesting that verticality intended in Gothic structures to raise the visitor's thoughts toward Heaven. The large supporting columns and arches along the nave are capped by rib vaulting. The floor is laid in a diamond pattern, and there are two rows, one above the other, of six individual stained glass panels featuring single figures. Not elaborately adorned within, as in say a Baroque cathedral, Westminster Abbey nonetheless has many shrines and tombs which are themselves richly colored. The organ installation is silver and gold. East of Westminster, in the area of London known as The City, the financial district, is the Baroque St. Paul's Cathedral.

St. Paul's Cathedral was not a confiscated Catholic structure but an Anglican church from the outset, although designed by Sir Christopher Wren after the manner of St. Peter's in Rome. The location in The City is on Ludgate Hill. It is, thus, the site of St. Paul's Cathedral that is at once its most striking architectural feature. This location results in the building being easily seen from almost any point in the city where the view is not obscured by other structures. The architecture is Baroque, but there is a distinct flavor of Neoclassicism especially in the entry façade.

Fig. 6-3 St. Martin in the Fields

A broad but shallow set of steps lead to the western main entrance which lies behind a portico formed of six paired Corinthian columns. Above these is a second story of four paired columns superposed by a Classical pediment. Flanking the two stories are square towers topped themselves by columned turrets. As in the case of St. Peter's in Rome, however, it is the central dome of St. Paul's that casts its distinct presence over London. Similar to the dome of St. Peter's and suggestive of the US Capitol, the structure is ringed with Corinthian columns and topped by a slender spire.

Of course, the western façade of St. Paul's is symmetrical, and its size is imposing. The dome is the second largest in the world after St. Peter's (http://www.sacred-destinations.com). Within, the arches along the sides of the nave are white with gold capitals, and the ceilings and the dome interior are richly adorned with paintings and gilt. The floor plan is the familiar Latin cross, and the floor itself is a black and white diamond pattern. The gallery just below the dome and above the nave is sometimes called "the whispering gallery," since a person at the wall on one side can be heard whispering by someone on the other side so geometrically perfect is the circle. West of St. Paul's on the eastern edge of Trafalgar Square is one last church to consider as compelling architecture in London: St. Martin-in-the-Fields.

Few locations are so auspicious as Trafalgar Square, standing as it does at the head of Whitehall. In addition to the square itself and its Nelson Memorial, there are the National Gallery of London and the National Portrait Gallery. The Church of St. Martin in the Fields stands on the northeastern side facing the square itself. The building is a simple rectangle with a Neoclassical western façade. The single entryway is a raised portico adorned with six Corinthian columns and pediment. Two columns stand on either side of the portico and more columns continue on each corner of the rectangle. The two stories of the church each have arched windows on both floors which extend in rhythmic repetition down each side.

Fig. 6-4 Salisbury Cathedral

While the church itself is not large by comparison to the cathedrals being examined here, its front and central tower and spire are and they point dramatically toward Heaven. Many Protestant churches in the US copied this architectural feature, therefore the tower and spire are familiar to American visitors. A square base with round windows on each side sits atop the church just behind the pediment. Above the base is a corner columned bell tower and surmounting this part of the structure is a spire with an octagonal base. The height and verticality of this tower and spire, widely adopted in the

US as noted, have undoubtedly come to suggest for many viewers a heavenly aspiration for God and are central symbols of the Protestant faith. Although it is a rather different one, it is the spire of the Cathedral at Salisbury, southwest of London, which also is its most compelling architectural element.

Fig. 6-5 Salisbury Interior

Salisbury Cathedral has several notable characteristics. Its spire is the tallest in Great Britain at 404 feet (http://www.salisburycathedral.co.uk). The walled cathedral close is the largest in the land at 80 acres, and the cloister is the largest in England. The clock in the aisle of the north nave is the oldest working clock in Europe and dates to 1386 CE. Remarkably, the cathedral was built in only 38 years, 1220-1258 CE, although it was another century before the spire was added. The western façade is one of the most highly decorated in either Great Britain or Europe with 67 statues gracing it. The stone used throughout, although not pure white, gleams as such in the sun.

Salisbury Cathedral is a beautiful and compelling example of architecture. It is neither on a hilltop nor centrally located in the small city of Salisbury, but its spire can be seen for miles around. Since the cathedral close is so large and walled, the unmistakable message is that here is a singular and most powerful place of worship. Not only the spire but the cathedral itself is of great scale. As noted above, the western façade is gracd with 67statucs. There are three peaked arches located symmetrically in the center of the façade, the middle one of which has a double entry into the building. Above these arches, there is a gallery of statues topped with three arched windows two stories in height. Along this central part of the façade are niches with statues one over the other, and at the corners are rows of such statues rising to small spires. Looming over everything is the central spire. The spire is not highly ornamented but gains power from its size.

Fig. 6-6 Bath Abbey

The interior of the cathedral is just as architecturally powerful as the exterior. First, its scale is enormous especially in the vertical dimension. Ranging along the sides of the nave and behind the choir are stained glass windows which illuminate and color the interior. Clustered Doric columns line the nave and are connected by rhythmically repeated peaked arches. Two more tiers of shorter columns and arches separated by windows rises above these, and the ceiling is ribbed vaulting. Back outside, the cloister is surrounded by an arcade with arches separated by small buttresses. The arches utilize the same clustered columns of the interior, and each has stone tracery featuring repeated geometrical patterns. Nearby in Bath, the Roman spa town, there is an abbey which in its own more humble way also represents compelling cathedral architecture.

While travelers no doubt make pilgrimages to Salisbury in order to see the cathedral, Bath Abbey is probably well down the list of reasons to visit Bath. The refurbished spas and the Georgian architecture of the Circus and the Royal Crescent usually lead the list. However, the Abbey is centrally located adjacent to the spas on a hill overlooking the River Avon. While not the highest location in Bath, this is certainly the city's cultural and tourist center. The abbey is a late example of the architectural style known as Perpendicular Gothic (http://www.sacred-destinations.com). The western façade has a central arched entry portal flanked by two slender towers. These towers feature ladders with figures climbing upward to Heaven and downward as well.

Above the entry is an enormous two story arch filled with stone tracery and stained glass windows after the manner of earlier rose windows but without the shape. Similar windows grace the three other sides of the Abbey and are separated by flying buttresses. At each corner of the exterior, as well as at each corner of the central tower, there are smaller towers with peaked tops. Today, the light-colored, so-called Bath stone is unfortunately coated with the soot of previous years' worth of smoke from residences and nearby industries.

Fig. 6-7 Fan Vaulting, Bath Abbey

Small by comparison to such houses of worship as Salisbury and York, the interior of Bath Abbey matches the larger ones with diffused light from its many stained glass windows and its fanciful fan vaulting. The nave is flanked with clustered Doric columns after the manner of Salisbury. At the eastern and western ends of the nave, stained glass windows depict scenes from Jesus's life (http://www.sacred-destinations .com). While the exterior of Bath Abbey is notable enough, it is the interior which is compelling architecturally and which is most likely to have a persuasive impact on the viewer. Just as there are a multitude of houses of worship which might profitably be studied in London, so there are many others in the countryside that might be considered here. Cathedrals at Wells, at Ely, and at Lincoln, not to mention Canterbury, would be worthy candidates. None, however, are likely to surpass the powerful effect of the Minster at York.

York Minster is the largest Gothic cathedral in northern Europe located within the walled city of York (http://www.yorkminster.org). Minster was the designation given in the Anglo-Saxon period to churches whose primary duty was missionary work, but York Minster is the seat of the Archbishop of York making it also a cathedral. As noted, the cathedral lies within the walls of an earlier, smaller York near the center of that old city. Its towers are sufficiently tall to be visible from virtually any part of present-day York as well as for some distance outside the city. One approaches it from the west along a cobbled street populated by small shops. Then it is just there in great mass and weight. It is also radiant in sunlight, giving an almost gold appearance with its magnesian limestone.

Fig. 6-8 York Minster

The minster has a Latin cross floor plan with a great stone tower above the crossing of the nave and transept. The western façade features two smaller towers flanking a central portal with double doors. Above the portal rises a two story stained glass window of delicate stone tracery topped with a peaked arch. Each of the flanking towers also has an arched entryway. There are four rows of superposed peaked alcoves all across the western façade, which at some point probably contained statuary. Alcoves like these continue on the north and south sides, but some of these do have statues of religious figures remaining in them. The arches, peaks, and tracery are all characteristically Gothic, and the towers each have ornate turrets at each corner and on each side. Each flying buttress has a similar turret. The southern façade has a large rose window at its peak.

Like Salisbury, the interior of York Minster is cavernous and both brightly and colorfully lit by 128 stained glass windows. Along the nave, clustered Corinthian columns are topped with peaked arches. Rising above these are two more levels of arched stained glass windows. The ribbed vault ceiling is white with periodic gold medallions at the intersections of the ribs (http://www.yorkminster.org). North of York is the old Norman or Romanesque Cathedral of Durham, built in memory of St. Cuthbert.

Durham is a very small city for such a prominent cathedral, but its history and geography, as well as those of the cathedral, are most interesting. The city itself is extremely hilly. Many residences, hotels, and hostels lie steeply uphill from the city center. Even the train station is uphill from the center. The city center is in a small depression below a ridge whereupon are located Durham Cathedral, Durham Castle, and Durham University. This location is on a peninsula of land created by a U-shaped bend in the River Wear. The river forms a natural moat for the castle, and the craggy cliffs atop which sit the three institutions provide additional defenses against the enemies of ancient times. Today, they merely make exploration of these places strenuous!

Fig. 6-9 Durham Cathedral

Although it is a Romanesque, or as the English call it Norman, piece of architecture, Durham Cathedral shares many primary features with York Minster. It has the same floor plan, the same three towers, and the same imposing presence. Since it is at the top of a hill, it has an even more compelling site. It is, however, much less ornate, inasmuch as there are none of the elaborate carvings and pinnacles of the Gothic style and no statuary on the exterior, although there are alcoves where some may have once stood. The western façade sits virtually on a gorge above the river and is no longer used as an entryway. Visitors enter through the north door just behind the western towers. The western towers are topped with simple pinnacles at each corner, but the large central tower over the crossing of the nave and the transept has none. Conical spires grace the corners of the eastern wing.

The interior of Durham Cathedral is the best evidence of Norman, rather than Gothic, style. Without the numerous stained glass windows of the Gothic style, although to be sure Durham does have some, the interior requires electric lighting to be anything but dim and foreboding. The interior also does not soar so high, and the columns are massive rather than delicate. That evidence of strength is clearly apparent and is perhaps the most distinctive architectural feature of the interior. One has the impression that this building might stand as long as has the Parthenon. While Durham Cathedral is the last of the English cathedrals to be considered here, there is one more house of worship left before leaving the United Kingdom for good and that is the Cathedral of St. Giles on the Royal Mile in Edinburgh, Scotland.

Fig. 6-10 Cathedral of St. Giles

Visitors to Edinburgh are presumably drawn to the Royal Mile, that street in the older part of the city leading from Edinburgh Castle at the top to the Palace of Holyroodhouse at the bottom. There are several notable historical stops along the route, but the most striking one architecturally is surely the Presbyterian Cathedral of St. Giles. The term cathedral is more traditional today than accurate, and it is a relic of the time before the Protestant Reformation. Since there are no bishops in the Scottish Presbyterian Church, the name of the structure is actually the High Kirk of St. Giles.

St. Giles might best be described as restrained Gothic. It clearly has both exterior and interior features that are noticeably Gothic, but massive size, huge towers, tall spires, and even large flying buttresses are missing. The western façade entryway is a single recessed portal without sculpted ornamentation, and along the façade are several small spires. The most dramatic architectural feature is the central tower over the crossing of the nave and transept, which is in the shape of a royal crown. The tower resembles a similar open one at the Victoria and Albert Museum in London. The floor plan is basically that of a Latin cross, and the interior is lined along the nave with widely spaced octagonal columns rising to and separated by rounded arches. The vaulting is ribbed. There is no rose window, but instead there are numerous vertical peaked stained glass windows which illuminate the interior.

Cathedrals and Churches in the United States of America

As recorded in a previous account of Protestant churches in America (Ragsdale 2007), the architecture of houses of worship across the Atlantic from the British Isles is extremely varied. One finds in many rural areas of the United States, for example, simple frame buildings which serve as churches. There are also converted barns housing "cowboy churches," and in many large cities there are worship services being conducted in converted store fronts. Moreover, America has frontier examples of Spanish mission architecture, such as the Alamo in San Antonio, Texas, and throughout the country there are to be found numerous examples of traditional Protestant church architecture often resembling London's Church of St. Martin-in-the-Fields. More than anything else, it is the diversity, and sometimes eclecticism, of church architecture that seems to characterize American houses of worship. Given the limitations of this work, it seems appropriate in the case of the United States to select just a few outstanding examples of church architecture as visual persuasion rather than trying to account for every style and type. For the present purposes, four houses of worship stand out: Washington, DC's National Cathedral, Manhattan's St. Patrick's Cathedral, Salt Lake City, Utah's Mormon Temple, and San Francisco's Grace Cathedral.

Fig. 6-11 National Cathedral

The nation's capitol is a treasure trove of significant architecture, including for example the Capitol, the many monuments and art museums, and the White House. Although the US Constitution makes it clear that the church and the state are separate governmentally, it is also true that Americans think of their country as a Christian one and one founded by those seeking religious freedom. As a result, it seemed only appropriate that some sort of "national" house of worship should be built in Washington. Additionally, it seemed only appropriate that this building should be a spectacular example of traditional church architecture. For this second purpose, the Gothic style was chosen.

The idea of a national cathedral for the United States dates to the time of the nation's first President, George Washington, who commissioned Pierre L'Enfant to design the city (http://www.nationalcathedral.org). A great national cathedral was L'Enfant's idea, but it was not until 1907 that the foundation stone was laid during the presidency of Theodore Roosevelt. The location on Mount St. Alban is the most commanding site in the District of Columbia. The building was not completed until 1990 during the presidency of George H. W. Bush. The official name of this Episcopalian house of worship is the Cathedral Church of St. Peter and St. Paul.

Fig. 6-12 St. Patrick's Cathedral

The design is latter-day Gothic, and the National Cathedral bears a strong resemblance to York Minster in northern England. The central tower stretches 30 stories toward Heaven. There are 231 stained glass windows, 112 gargoyles, and 288 angels. It is the sixth largest cathedral in the world and, behind the Cathedral Church of St. John the Divine in Manhattan, the second largest in America (http://www.nationalcathedral.org). Undoubtedly because of its relative newness, the cathedral still sparkles in the sunlight with its off-white stonework.

While its site, its size, and its height are its most important architectural features, the National Cathedral has all of the usual visually compelling Gothic features. The western façade features three arched and recessed portals, each with two entryways. As in the case of most of the French Gothic cathedrals, portal sculptures offer Bible stories. The entry façade is beautifully symmetrical with the central portal topped by a large rose window. The corners of the western entrance feature twin towers with Gothic pinnacles on each corner. The central tower is located over the crossing of the nave and the transept. Along the remaining three sides range slender, but large, flying buttresses. Inside, the National Cathedral has the usual cruciform floor plan. Large, widely spaced clustered columns line the nave and point upward to a rib vaulted ceiling.

Perhaps the most visually compelling Roman Catholic house of worship in the United States is St. Patrick's Cathedral in midtown Manhattan. The location is prime real estate in midtown Manhattan, and it is the site or location that is one of the main architectural elements in the visual persuasiveness of St. Patrick's. The cathedral was begun in 1858 and completed in 1908 (Costecalde 1997). Like the National Cathedral, St. Patrick's is a Gothic edifice, but in contrast to some English and European cathedrals it has a comparatively narrow western façade. Unlike, say, Notre-Dame de Paris, St. Patrick's has two conical spires instead of towers on the north and south corners of the western entryway. Three deeply recessed portals grace the entryway, and each is topped with a triangular peaked arch. Above the central portal is a large rose window. The triangle theme is repeated in the stories above, and in conjunction with the twin spires gives a strong vertical impression. This theme of triangles and spires is repeated on the north and south side buttresses as well as entries.

Clustered Corinthian columns, widely spaced, line the nave within and are separated by peaked arches. The ceiling is ribbed vaulting. Looking back over the entryway, the stained glass of the rose window has angels as its theme, with the eight rose petals symbolizing Jesus's eight beatitudes given in the Sermon on the Mount (Costecalde 1997). Just below this window is the Great Organ with 7,855 pipes. The floor plan is the traditional Latin cross. There are many churches in New York City, but none surpass St. Patrick's Cathedral in compelling form.

Salt Lake City, Utah, is the home of the religious group known as the Church of Jesus Christ of Latter Day Saints, commonly known as the Mormon church. This religious group was founded by Joseph Smith in New York, but persecution caused large numbers of Mormons to migrate to the more sparsely settled areas of the American West. They were led by Brigham Young and finally came to stay on the edge of the Great Salt Lake in Utah. The location, a valley ringed by snow covered mountain tops and vivid blue skies, is one of spectacular natural beauty. Within Salt Lake City itself, there is an elevated portion in the very center where the Mormons have built Temple Square. The square is visible from many downtown locations and includes the Mormon Tabernacle and Temple. It is the Temple which is of architectural interest here.

Fig. 6-13 Mormon Temple

The Temple does not fall readily into an architectural style, although it incoporates many traditional elements of church architecture. Gothic elements are the most noticeable, but there is very little of the exterior ornamentation associated with Gothic design. That which is present represents figures and beliefs unique to Mormonism, such as a statue of the angel Moroni. The orientation of the Temple is toward Jerusalem and Solomon's Temple. The eastern façade has three towers topped by a short square base spire. The towers are ornamented with simple arched alcoves one above the other. The western end of the Temple repeats this design, but the height of the towers and spires is less than those on the east. Along the sides, there are two rows of simple arched windows and two interspersed rows of smaller round windows. The interior is regarded as sacred by Mormons and access is, with few exceptions, limited to members of the faith and not every one of those.

At the top of Nob Hill overlooking the city of San Francisco, California, stands Grace Cathedral, the seat of the Episcopal Diocese of California. The hillside location of the cathedral provides a site of visual prominence and exemplifies the use of architectural elements to establish influence. The architectural design is Gothic, and the entry façade bears no small resemblance to Notre Dame de Paris although not so grandiose. Twin towers rise above the entrance and a spire is erected at the crossing of the nave and the transept. Notably, the entrance features replicas of Lorenzo Ghiberti's bronze baptistery doors depicting the Gates of Paradise. Above the doors is a typical rose window, and the interior is replete with stained glass. The visual power of Grace Cathedral is clearly drawn from its Gothic architecture and its hilltop site.

Fig. 6-14 Grace Cathedral

This account of only four of the thousands of houses of worship in the United States has not been an effort to exemplify American church architecture. Rather, it has been an account of a few of the best known and most architecturally compelling American houses of worship. To complete this assessment of the visual persuasiveness of houses of worship in the English-speaking world requires a continued venture westward, this time across the Pacific Ocean to Australia and New Zealand. As in the case of the United States, architecturally distinctive houses of worship will be the focus. Two churches each in Australia's two largest cities, Sydney and Melbourne, are the beginning points. Then, two churches in New

Zealand's capital of Wellington, and one in the largest city of the South Island, Christchurch, will conclude the account.

Cathedrals and Churches in Australia and New Zealand

On the eastern edge of Hyde Park in Sydney's city centre stands majestic Roman Catholic St. Mary's Cathedral. Along with a large stonework plaza in front of its entrance, this site is a propitious one which makes the cathedral stand out visually. There are no buildings clustered near it to detract from the view, and the nearby parks serve to frame the building in greenery. The architecture is Gothic Revival featuring twin spires on either side of the main entry portal and a tower rising over the crossing of the nave and the transept. Unlike the usual pattern for Gothic cathedrals, the entry faces south rather than west, otherwise St. Mary's resembles the western façade of Notre Dame de Paris. The interior is richly illuminated by stained glass windows and features ribbed stone vaulting. The structure is the largest cathedral in Australia.

Fig. 6-15 St. Mary's Cathedral, Sydney

St. Andrew's Cathedral, Sydney, is a much smaller structure than St. Mary's. Also located in the city center, St. Andrew's is the seat of the Anglican Diocese of Sydney. Unlike St. Mary's, it is surrounded by many tall buildings, making its site much less visually impressive than St. Mary's. However, the building is nonetheless visually compelling. The architectural style is Perpendicular Gothic, which maximizes the visual power of what is otherwise a rather compact building. The cathedral's western entrance is suggestive of York Minster, with two towers rising at each side. A smaller tower rises above the crossing of the nave and the transept. Each of these three towers is decorated on its four corners with the curlicue pinnacles of Gothic style. The interior features stained glass by Hardman and Company of Birmingham, England, which company also was responsible for the windows of St. Mary's.

Fig. 6-16 St. Andrew's Cathedral, Sydney

St. Patrick's Cathedral in Melbourne, Australia, is the seat of the Archdiocese of Melbourne. Its architecture is Gothic Revival, meaning that its principal features are Gothic but the structure does not share all of the traditional English, French, or German Gothic features. It is located on one of Melbourne's hills and surrounded by enough lawn to keep it from seeming hemmed in. Its site is additionally prominent and architecturally compelling inasmuch as the cathedral is located behind Melbourne's Parliament of Victoria building.

Fig. 6-17 St. Patrick's Cathedral, Melbourne

Unlike the majority of Gothic cathedrals, St. Patrick's faces south and is virtually devoid of sculptural ornamentation. There is a single entry portal, which is recessed under a peaked arch. Above is a tall, multifaceted window, which includes a rose at the top and is arched like the portal. On either side of the entryway are bell towers culminating in multi-sided spires. Behind these spires, where the nave and the transept cross, is a similar but much taller central spire. While the exterior architecture in general is visually compelling, the stonework is dark and unattractive, giving an austere impression. Within the cathedral, is a comparatively narrow nave lined with clustered columns separated by peaked arches. There is a spare usage of stained glass, especially in the clerestory windows behind the altar. St. Patrick's Cathedral can be seen easily from many parts of Melbourne and thus stands as a powerful symbol of the faith in the city.

St. Paul's Cathedral in Melbourne is the seat of the Anglican archdiocese of Melbourne and sits not on a hill but near the banks of the

Yarra River at the foot of Swanston Street. This location is clearly the commercial and cultural center of the city. Across one street is Flinders Street Station, Melbourne's iconic railway station. Across another is Federation Square, with its art museums, restaurants, and shops. The cathedral's site, therefore, is just as impressive as that of St. Patrick's albeit in a different way. While not so large as St. Patrick's either, St. Paul's easily figures prominently in one's view in many downtown locations.

Fig. 6-18 St. Paul's Cathedral, Melbourne

St. Paul's Cathedral is Gothic Revival as well, and the structure shares many features with St. Patrick's. Even so, the two cathedrals are quite distinctive visually. St. Paul's also has a southward facing main entrance, twin towers with spires flanking a central arched portal, and a massive central tower. The stonework of the building is dark interspersed with beige stones, and the spires are burnt sienna. The towers feature more windows and arches than St. Patrick's, and there are Gothic finials at each tower corner. Inside, stained glass behind the altar and along the sides provide colorful illumination. Clustered columns line the nave separated by peaked arches, but these columns are made of beige marble with dark gray accents. The floor features a variety of varicolored geometric patterns. There are, of course, other churches and styles of architecture to be seen in Melbourne, but St. Patrick's and St. Paul's are probably the most representative from the standpoint of visual persuasion. Across "the ditch," as New Zealanders call the Tasman Sea separating them from Australia, are two more cities with visually distinctive architecture. The first, located on the North Island, is Wellington, the nation's capital, and the second, located on the South Island, is Christchurch.

There are two architecturally significant houses of worship in the comparatively small city of Wellington. The first is the recently completed Wellington Cathedral of St. Paul, which is the seat of the Anglican archdiocese of Wellington. While this is a handsome building, it is not so architecturally interesting as the much older Cathedral Church of St. Paul's (Old St. Paul's) nearby. This latter structure is no longer a parish church, and it is now under the care of a historical trust, but it continues to be used for funerals, weddings, christenings, and other such ceremonies. Both churches stand on the hill which also includes Parliament and associated governmental buildings. For this reason, both houses of worship have auspicious sites, although the newer and taller building is much more prominent than the older one.

Fig. 6-19 Wellington Cathedral of St. Paul

Wellington Cathedral is a simple rectangular basilica with two stone pillars flanking the single entryway on the southern side and a massive bell tower on the east side where a central spire might stand were there a Latin cross floor plan. Between the two stone pillars is a tall, slim window with a cross superimposed thereon. The building material is reinforced concrete used to make the structure more resistant to the earthquakes to which this island nation is prone. The coloring of the building is primarily burnt sienna. Within, there is good illumination provided by numerous modern stained glass windows depicting scenes from the life of St. Paul. Behind the altar is a mosaic of the Madonna.

While Wellington Cathedral is architecturally compelling in its own right, it is Old St. Paul's nearby which is a uniquely satisfying example of the visual power of architecture. This cathedral church is done in a Gothic Revival style, but the real surprise is that it is constructed entirely of native New Zealand timbers down to the nails. Aside from the trim work and the steeple, Old St. Paul's is painted white. The building is comparatively small and has the simple rectangular basilica form of the newer Wellington Cathedral. The entrance façade faces west and consists of three portals with simple triangular peaked arches. Above the central portal is the steeple. The interior is ringed with colorful stained glass and carpeted in maroon, which together impart pleasant warmth. There is an extensive use of arched and exposed beams of medium brown timber, which rises to a peaked ceiling.

Fig. 6-20 Old St. Paul's Cathedral, Wellington

ChristChurch Cathedral in the South Island city of Christchurch is the final stop on this global tour of visually persuasive architecture. Over the years since its original construction, the cathedral has been damaged by the earthquakes common to this island nation. Regrettably, ChristChurch Cathedral has again been seriously affected by the magnitude 6.3 earthquake which struck Christchurch on February 22, 2011, destroying the church's spire. This account is based on the author's visit to Christchurch in July, 2010. At this writing, the cathedral is still unsafe to enter, and plans are afoot to raise the funds necessary either to rebuild it as it was or to raise a new building altogether. For those concerned about visual prominence, the fervent desire is that the cathedral be rebuilt as it was.

Fig. 6-21 ChristChurch Cathedral

The Anglican ChristChurch Cathedral is visually distinct from all the cathedrals discussed in this chapter. While its Gothic Revival style is not unusual, the use of stonework is. The majority of the stonework is dark in color, while the windows, doors, and tracery are light. There is a cruciform floor plan, and a single tower and spire stand to one's left facing the entrance. Although the cathedral is not unusually large, the dark colors and

Gothic design make it appear quite massive. The design, while Gothic, is not excessively ornamented, rather it is quite modest in adornment. The site of the structure is the commercial center of Christchurch fronting Cathedral Square. The building is set off by the open space before its entrance.

As in the case of the cathedrals of Continental Europe, the visual persuasiveness of these buildings in the English tradition is rooted in such fundamental architectural elements as line, rhythm, scale, space, mass, and, most importantly, site, light, color, and ornament. Like their Continental counterparts, these churches and cathedrals hold forth the belief of their builders that the house of God should symbolize the nature of Heaven, should be ornamented in glass and stone with representations of biblical messages, and should stand alone as a mute example of God's power and majesty.

CHAPTER SEVEN

UNIVERSITIES AS VISUAL PERSUASION

When considering universities, one's first thought is likely to be about higher education. Even more so than in the case of museums of art, universities do not spring to mind when one thinks of structures as visual persuasion. It is therefore the first order of business in this chapter, to make the case for universities as means of visual persuasion. While the primary mission of a university is education, there is much more that is involved. While universities are known as institutions of learning, their buildings are anything but institutional in the strict sense of the term. At this writing, universities, especially in the US, are gradually shifting away from the brick and mortar venues which have for centuries defined campuses in the popular mind by offering many of their courses online. Although online education is merely a technologically advanced version of the familiar correspondence courses of the past, it offers improvements over that earlier experience and often provides a cheaper alternative to campus residence. What online education does not, indeed cannot, do is visually to influence parents, students, alumni, faculty, support personnel, and those who support the university financially in a compelling manner.

There are, however, unquestionably important general social influences which universities have on a wide range of stakeholders. Let us begin with the most significant of these: the university as a persuasive force in a potential student's decision to attend. At the author's place of employment, Sam Houston State University in Huntsville, Texas, USA, a new President arrived in 2001. This man, James F. Gaertner, was a graduate of SHSU and a native Texan, both important ingredients in being selected a university president in this state. Dr. Gaertner proclaimed immediately that during his tenure—he is about to retire at this writing—"academics is first." Such a declaration is what nearly every faculty member, if not every student or alumnus, wants to hear. However, Gaertner went about implementing this slogan in what some considered an unusual way: he launched the first major construction program in many years. How was such a plan a matter of academics first? The logic is sound if a little complex.

Prior to Dr. Gaertner's arrival, at least two things were quite apparent at SHSU. The first was a rather stagnant rate of growth in enrollment and an off-campus physical environment that was, to put it bluntly, ugly. Gaertner understood a fundamental principle of the visual persuasiveness of architecture. He realized that there must be an attractive place for them to come to if one wished to entice students. Students and their parents certainly weigh many factors, such as cost, geographic proximity, and reputation, in the decision to choose a university. However, no one would choose to live in an ugly place if there were more attractive alternatives. Notably, the university in short order bought property adjacent to the existing campus, which was populated by empty buildings, bars, restaurants, and other quite dilapidated and neglected structures and immediately razed it. In its place, the university erected a large, traditionally styled dormitory complex for freshmen. While the area is not to this day devoid of architectural eyesores, the general effect was to transform the appearance of the campus along its main border. Now, no parent need wonder whether or not a student might be lured into a seedy and dangerous place.

Dr. Gaertner also razed on-campus dormitories and replaced them with fresh and attractive ones. A parking garage, a first for this campus, was built, the student union building was remodeled, a set of former dormitories converted to offices was razed and replaced by an open landscape of concrete and plantings, and a new building for the College of Humanities and Social Sciences was added among other things. As Gaertner prepares to depart, a large performing arts complex is near completion, and the university envisions a School of Nursing complex and possibly an engineering building. Although not an unattractive campus at the outset of Dr. Gaertner's presidency, today's SHSU is a major transformation of visual power. To be sure, there have been many ingredients to it, not just architectural ones, but today's enrollment reflects a steady increase since this president took office. For Gaertner, improving the university campus architecturally meant growth in enrollment, a greater retention rate, increased entrance requirements, and as a result placed academics first.

In addition to attracting students, the visual power of architecture on a university campus also serves to influence the experience of students during their college years. Again, many other ingredients also influence this experience, but the living campus environment contributes in several ways to the years of learning. One of the most obvious of these ways is in providing a favorable environment or context for study, contemplation, and enjoyment. Classrooms should be well-lit, comfortable, and have good acoustics. They should be housed in buildings that are visually attractive if not compelling. There should be easy access between buildings and space for students to congregate, to visit, and to hold meetings both inside and outside buildings. There should also be green open spaces with attractive plantings and trees, making a park-like environment that is enjoyable to inhabit.

Inevitably, students who obtained degrees from architecturally compelling campuses remember the favored old spots long after their graduation. The effect of architecture can endure in a former student's memory remarkably well. At SHSU, an administration building with classrooms known as Old Main burned to the foundations a number of years ago. Rather than either clear the spot or rebuild the building, SHSU chose to leave the lowest part of the walls standing surrounding the remaining interior stone stairs and what is now a grassy floor. Not only is this spot a frequent gathering place for students, it is also a mecca for former students. President Gaertner, for example, can remember the location of the classroom in the building where he took a history class as an undergraduate.

This campus environment not only profoundly affects students and continues to do so far after graduation, it also affects faculty and staff. Some schools may have enough prestige to attract faculty without regard to environment, but most need to provide the same kind of living and working space for faculty they wish to attract and retain. The author's undergraduate alma mater, Samford University in Birmingham, Alabama, USA, was known as Howard College for many years and was located in the East Lake district of the city until a completely new campus was begun which opened for the Fall semester, 1957. The old campus, for whatever nostalgia there might have been for it, had not only been outgrown, but was shabby and in disrepair. To many potential students, the campus, indeed the college, would be an embarrassment. For over 50 years now, however, the new campus, set on a hill in a beautiful valley and with a uniform Georgian Colonial architecture, has been a place that stakeholders are eager to represent.

A university is also a cultural icon, much in the same way a museum is. Certainly great museums lend status to the cities, states, and countries in which they are located. They are a mark of sophistication, taste, and wealth. Even lesser museums mark a place as gentile and not just commercial and industrial. So it is also the case that the locating of a university in a city or town lends the same sort of prestige. There is also normally a deliberate connection where possible between a university and the city's arts complex as reflected in the phrase "town and gown."

Huntsville, Texas, would hardly be the place it is today without its university any more than would Oxford or Cambridge in England or even Paris without the Sorbonne. It is time, then, to begin an assessment of the visual persuasiveness of architecture to be found at universities. There are too many, of course, for the approach to be exhaustive any more than the assessment of structures in previous chapters has been. As in those previous chapters, a few of the world's most distinctive universities will be considered along with a number of others which represent different trends. The beginning point is with the two English universities which for many represent the quintessence of the idea of a university: Oxford and Cambridge.

Oxford University

Americans especially tend to think of universities as being located on campuses. In this sense, campuses are large tracts of land devoted entirely to the variety of structures which constitute the university: classroom buildings, administration buildings, student centers, dormitories, laboratories, and the like. Campuses are ordinarily situated within cities but typically are set apart as units. Of course, there are exceptions to this notion of a campus. Some city universities are contained within single high rise buildings, as is the case for example with Roosevelt University in Chicago, Illinois, and this arrangement is also to be found in other institutions around the world. However, both Oxford and Cambridge have a rather atypical arrangement. Both are constituted by many small individual campuses spread throughout the cities in which they are located. These campuses, known as colleges, are self-contained and include living quarters, study halls, eating establishments, classrooms, and houses of worship. The colleges are self-governed but ever mindful of the overall imprimatur of the entire collection of them, the university itself. American universities are also made up of colleges, but these are usually administrative units rather than separate entities and are not self-contained.

What the British idea of a university, as expressed in the cases of Oxford and Cambridge, means for this chapter is that each college has a separate visual character. Each looks different and impacts the viewer with varying visual power. In some cases, individual buildings are especially architecturally compelling, but it is the college as a whole which illustrates the idea of compelling form. With one type of exception, this chapter will examine several of the many colleges at Oxford. The exception lies with the shared buildings of the university, such as the Bodleian Library, the Sheldonian Theatre, and the Ashmolean Museum.

Oxford University is spread throughout the city of Oxford in a way that suggests the city built up around and between the colleges. To be sure, there is an area of Oxford that consists almost entirely of colleges, but one is never far from commercial buildings, banks, and restaurants. As in the case of virtually every university, this free access to both town and gown is undoubtedly a strong attraction for potential students. To be sure, the academic reputation of Oxford is probably the principal attraction for students, but the idea of spending four or more years in such a handy location may be very powerful by itself.

The colleges themselves are typically walled or fenced areas with an inner lawn or yard surrounded by living quarters, classrooms, and a chapel or church. Trinity College and All Soul's are good examples of the interior yard, while Merton, Magdalen, and Christ Church Colleges illustrate a variety of walled spaces. Many of the college buildings and chapels at Oxford are heavily influenced by the Gothic style, with its ornate towers, decorated portals, and intricate spires. On the other hand, common structures such as the Ashmolean Museum and the Sheldonian Theatre are Neoclassical in appearance. Merton College is a typical example of the Gothic style. This is especially the case with Merton Chapel, which is a small Gothic cathedral.

Fig. 7-1 All Soul's College

Fig. 7-2 Trinity College

Merton College was founded in 1264 CE and was the first self-governing college at Oxford (http://www.merton.ox.ac.uk). The stone walled college sits adjacent to its large playing field used for football (soccer) and other athletic pursuits. Its gated entrance off a small lane is a single portal within a recessed arched entryway. The yellowish stone wall features paired ornate windows, dormers, and chimneys, and these alternate in a rhythmic pattern all around the buildings. The entry is bounded on either side by pillars. Above it is a small central frieze on either side of which is a peaked alcove with statuary.

Merton Chapel is considerably more ornate than the remainder of the college, although it complements the other buildings very well. Its entryway is similar to the college entrance. There is a single portal within a plain stone façade topped by a massive double sectioned window with two rows of seven panes defined by scrolled stone tracery. On either side of the window are alcoves with statues. The roof above is peaked. Square turrets with Gothic spires are at each corner of the entry façade. On either side of the entry projection are smaller versions of the decorative window on the front. Rising behind the entry to the chapel is a central tower topped with Gothic spires at each corner and side. At the rear of the chapel is another, albeit smaller, ornate window with seven panes. There are north and south transepts along with a nave and two side aisles.

Fig. 7-3 Merton College

Of the common buildings at Oxford, four are particular compelling pieces of architecture. One of these, the Sheldonian Theatre was examined in Chapter Four. The Bodleian Library and its reading room, Radcliffe Camera, represent two different architectural styles. The main library itself is a foursquare structure with an inner courtyard and design elements reflecting the Gothic style, while the freestanding circular Radcliffe Camera is largely Neoclassical. The main building of the library has three stories above ground. Its walls, like those at Merton College, are largely stone without ornamentation other than windows. Along the top of the walls are rectangular battlements interspersed with Gothic turrets. Above the main entrance rises a tower with similar Gothic turrets at each corner. The library is neither colorful nor heavily ornamented, thus the primary visual impact is gained through its mass. In fact, it resembles a fortress or even a prison to a large degree.

Fig. 7-4 Bodleian Library

Adjacent to the Bodleian main building is Radcliffe Camera. The library is for the most part a reference only institution. Students and authorized others may request documents for study, and when these documents are delivered it is to the Radcliffe Camera reading room. This structure is circular in design and adorned with Neoclassical elements. Like the library, the building has three above ground stories. The ground floor is rough stone work and has entries interspersed with alcoves ringing the base. The second and third stories are of smooth stone of variegated color ringed with paired Ionic columns. Windows replace the first floor's entries and alternate with empty alcoves. Atop the three stories is a balustrade encircling the building and a central dome topped with a small turret. While the effect may not have been intentional, Radcliffe Camera is reminiscent of the Pantheon in Rome save for the missing portico.

Fig. 7-5 Radcliffe Camera

The Ashmolean Museum, located some distance away from the central cluster of colleges at Oxford, is an example of traditional Neoclassical architecture with its Ionic columns and pediment at the main entrance. It is the oldest art museum in the United Kingdom and one of the oldest in Europe, having been established in 1677 with the gift of his collections by Elias Ashmole (http://www.answers.com). Although not so large, the Ashmolean Museum resembles the British Museum discussed in Chapter Three. There is a central building flanked by projecting wings. The main entryway features four fluted Ionic columns and one more on each side. Above the columns is a typical Greek pediment. The wings on each side feature smooth Ionic columns, and this theme continues on an adjacent street side of the museum.

At Oxford University, a number of architectural elements come together to produce an impressive and visually compelling experience. The site is a city itself, with the numerous colleges interspersed throughout the location. Each college stands behind a walled courtyard of lush greenery and lawn, and each is designed in a harmonious and complementary architectural style with Gothic dominating. The chapels resemble small cathedrals, and buildings such as the Bodleian Library are massive in size. The visitor has the impression of desirable living, the fostering of contemplation and learning, and the power of knowledge itself. It is a place that would undoubtedly be a memorable one to have experienced for four or more years. Northeast of London about an hour's travel by train is the other British educational icon: Cambridge University.

Fig. 7-6 Ashmolean Museum

Cambridge University

Oxford and Cambridge are well-known as rivals in almost every field of endeavor, from sports and debate competition to a rivalry for Nobel Prizes. From a visual persuasion point of view, however, the two universities are more alike than different. Certainly, one is not more beautiful or more visually compelling than the other. Like Oxford, Cambridge offers an extremely supportive and perhaps idyllic environment for students to learn and for academics to teach and research. Cambridge seems to be a bit more compact than Oxford, but the city and the university blend together just as they do in Oxford. Also like Oxford, Cambridge makes a considerable effort to incorporate green spaces, such as lawns and courtyards, and there is a very large park known as Christ's Piece to the east of the principal buildings for picnics and casual sports.

Queen's College and St. John's College provide good examples of courtyards.

Fig. 7-7 King's College Chapel

Perhaps the most distinctive and certainly the most visually compelling building at Cambridge University is King's College Chapel. This beautiful Gothic chapel is virtually the size of a cathedral, and its spires can be seen from many points on the campus. Its location is both central and across from several commercial establishments. The single portal entryway to the chapel is flanked by two slender towers topped by ornate Gothic turrets. Between these towers is a massive arched window with two rows of nine panes of glass. The slender tower theme is repeated along the sides of the chapel and along the wall surrounding the Great Court. The gatehouse of the Great Court is also marked by two slender towers matching those of the chapel. Between these two, however, is a single domed clock tower.

Further up the street from King's College is another visually impressive Gothic chapel, that of St. John's College, which is marked by a square tower with spires at each corner. Between King's College and St. John's is the Cambridge University Senate House used for graduation ceremonies and other special occasions. Unlike many of the rest of the university's buildings, this one is Neoclassical rather than Gothic. The Senate House is a two story structure of brilliant white stone. It is perfectly symmetrical with a single door standing in the center of four Corinthian columns under a Greek pediment. Continuing along the front in unbroken rhythm are four pilasters with Corinthian capitols and separated by two rows of arched windows all of the same size. The roof is ringed by a balcony.

Fig. 7-8 Senate House

What is remarkable about both Oxford and Cambridge is how completely and flawlessly they embody the idea of an entity—in this case a university—which acknowledges the importance of compelling form in attracting and developing students and faculty in a memorable way and in a way that puts beauty at the center of the image of a university. It is easy to see how the typical American university campus evolved from the English notion of a university as expressed at Oxford and Cambridge. This evolution is perhaps best evident at America's oldest university, which is Harvard University in Cambridge, Massachusetts. Before examining Harvard, however, there are three other universities in the United Kingdom worthy of attention, as well as four continental European ones.

Other Universities of the United Kingdom

London is also a city of universities. It is the home of the University of London, the Imperial College of Science and Technology, and the London School of Economics, along with many others. Just as the city itself is vast, so these institutions are also and spread out over the city in numerous locations. It is beyond the scope of this assessment to do justice to London's universities in their entirety. Their buildings represent several different architectural styles, most of which are extremely well-integrated into the other architectural forms surrounding them. London, like other national capitols, is a city of monumental buildings with grand architecture to match. A notable example of this architecture of grandeur in an educational institution is the Imperial College of Science and Technology, located near the Royal Albert Hall in the South Kensington area of the city.

Fig. 7-9 Imperial College

It has been noted often before that site or location is a recurring element of architectural influence, and so it is also true of Imperial College. Also nearby are such important structures as the Victoria and

Albert Museum, the Albert Memorial, the Natural History Museum, and Harrod's Department Store. All of these are examples in their own right of architecture used as visual persuasion. Additionally, South Kensington is a particularly desirable residential location as well as the location of Kensington Palace. A college or university located in such a place already has a basic architectural presence, but it surely also must fit its surroundings. Imperial College does. Its main building is a massive and ornate example of Neoclassical design in grayish white stone. It complements the other monumental buildings in the area and takes second place to none of them. The main entrance is a three story archway in the shape of an alcove. Flanking the stairs on the ground floor is ornate statuary and sculpted busts. The front of the building is decorated with a rhythmic procession of Ionic pilasters. At the top, there is an elegant balustrade. In addition to its location, then, Imperial College derives its archtitetural power from its mass, it ornamentation, and its use of the lines, angles, curves, and geometric forms of Neoclassicism.

Durham in the north of England has already been described as the home of the Norman Durham Cathedral. However, on the same hilltop location as the cathedral, there is also a thriving modern university. Most interestingly, from an architectural standpoint, University College of the University of Durham is housed in what was formerly Durham Castle. Other university buildings, such as the one housing the Department of Music, are also located nearby. Like Oxford and Cambridge, more university structures are located at other sites in the small city.

Durham Castle is situated across from the cathedral green on a promontory overlooking the River Wear. Of course, this location is a commanding one as was necessitated by the purpose of a structure built to withstand invading forces. The walls of the fortress were not designed with visual attractiveness in mind, but they are nonetheless compelling. The walls are topped by battlements and towers, and there is an octagonal castle keep. The entryway to the castle today is a wide arched passageway into an interior courtyard. The entryway is flanked by a central tower with both an arched and a round window. Within the walls, there are structures with domed corner and side towers.

Durham Castle is not ornate, or graceful, or colorful. Its architectural power derives from its hilltop location and its rugged mass. Castles and other similar structures are the subject of a subsequent chapter in this book, therefore only a few characteristics will be noted here. What is the visual experience, and is the visual experience significant for a university? Looking at the castle from afar, one certainly has a first impression of strength and power deriving from the solidity and texture of both the outer walls and the interior buildings. Geometric forms, such as squares and octagons, are used throughout and result in a sense of appropriate architectural form. If castles were intended to suggest that an enemy rethink plans to attack them, then certainly Durham Castle succeeds in that intention. For a student—living quarters are also incorporated within—there must be a sense of well-being in residing and learning in a particularly secure environment and one which would be almost uniquely memorable.

Fig. 7-10 Durham University

Edinburgh not only is home to St. Giles's Cathedral, examined earlier, and Edinburgh Castle, to be studied later, it is also the location of the University of Edinburgh. This university is also spread out over Edinburgh, although its heart, the Old College and a number of smaller buildings, is located primarily in an area of Edinburgh just off the Royal Mile. The Royal Mile is probably the most important real estate in this capital city, thus the university's location is its primary architectural feature. Edinburgh's architecture is often not especially graceful or delicate, and the buildings of the university are no exception. They derive their architectural influence, their visual persuasiveness, from their mass, their texture, and their strength, much like Durham University does.

Fig. 7-11 University of Edinburgh Old College

Old College has a street entrance with a Neoclassical façade. This portal for both pedestrians and automobiles leads to a crushed rock courtyard where one finds benches to rest surrounded by three two story buildings attached together by corner arcades. The ground floor of the courtyard is of rough stone with arched entrances and windows. Above this floor the façade is decorated with Corinthian pilasters. The corner arcades are ornamented with Ionic columns. There is a balcony surrounding the roof. The street entrance, a single portal, is ornamented with Doric columns and topped with a dome which uses Ionic columns. While the buildings of Edinburgh in general are marred by the dark coating of soot from the city's numerous chimneys, the visual effect is less impaired than it might have been had the buildings followed a more graceful and delicate style.

Continental European Universities

Rather than attempt either to catalog the universities of Continental Europe or to assess any single one in complete detail, it is the purpose of this part of the chapter to give a glimpse of the patterns to be found in a few of the most important cities of the continent. Paris is a city of architecturally compelling monuments, bridges, museums, and libraries, so it is no surprise that it is also home to one of the most storied names in higher education—the Sorbonne. Naturally, la Sorbonne is located in Paris's *Quartier Latin*, since the area was called the Latin Quarter because of the students there studying Latin. Like the English universities, the Sorbonne is not just one institution but several, each emphasizing one or more specific aspects of learning.

Fig. 7-12 Sorbonne

The building known as the Sorbonne is on one of Paris's grand streets, the Boulevard St. Michel. It sits back from the street at the end of a small square lined on each side by bistros and stores. The square rises a bit toward the building to give it some prominence. Its entry façade is a two story classical French Baroque design common to many Parisian buildings. There is a two story, single portal entryway flanked by six Corinthian columns. Above is a central façade with Corinthian pilasters and sculpture in niches topped by a plain Greek pediment. Characteristically for the French Baroque style, there is a large dome rising above all much in the manner of Les Invalides not far away and of St. Peter's in Rome. To say that the architecture of the Sorbonne is similar to other structures in Paris is in no way meant to minimize the architectural power of the building, and there would seem to be little question that the university's many students over the years have been strongly influenced by its beauty.

Humboldt University is Berlin's oldest university, and its main building is located at 6 Unter den Linden (http://www.hu-berlin.de). For those familiar with Germany's, and especially Berlin's, wartime history, this location was in East Berlin after World War II and behind the Berlin Wall during the Cold War. It has only been since the fall of the Soviet Union and the destruction of the Wall that visitors from the West have had ready access to the university and such other important architectural sites as Museumsinsel, the German History Museum, and the State Opera House.

Fig. 7-13 Humboldt University

In addition to its prime location on a beautiful avenue now restored to its pre-war state, Humboldt University is an impressive architectural sight on its own. Its entrance lies at the rear of an iron and stone fenced paved courtyard. Statuary adorns the stonework. The building has a main central entryway with two wings on either side which form the sides of the courtyard. The entryway on the ground floor to the three story structure is simple white stone with rectangular windows. The second story, however, features six Corinthian columns, which are two stories high. The pediment is unadorned, but it is topped with six metal statues. The second story windows are arched, while those on the third story are square. Like many of the other buildings on Unter den Linden, Humboldt University is a striking example of the visual power of architecture. With this very brief glimpse of two important Continental European universities, the next location of a university of compelling visual influence is in the US. It is Harvard University in Cambridge, Massachusetts, the oldest university in America and perhaps the quintessential American university. It is noteworthy that the city which has grown up around Harvard is named after Cambridge, England.

Harvard University

Harvard University takes much of its character and its architectural presence from Cambridge. Like Cambridge and a bit less like Oxford, Harvard is comparatively compact rather than spread throughout the city and is likely to have become the model for later US private and public universities. Where the colleges of Oxford and Cambridge are each set off as walled entities, Harvard is surrounded by a brick wall with several iron gates at strategic points. Unlike the English universities, the campus is readily accessible by the public.

Fig. 7-14 Harvard Yard

Also unlike Oxford and Cambridge, and for that matter most US universities, much of Harvard's architecture is quite uniform, giving it an essential look. The surrounding walls and the buildings in the oldest and most central area of the campus, Harvard Yard, are of dark red brick. Harvard itself dates to 1636 (http://www.harvard.edu/about), so there has been considerable construction outside the central part of the campus that is anything but consistent with the essential character of the older buildings. Some of these newer buildings, however, represent the very best of 20th century architectural style and will be considered here along with the central ones.

The heart of Harvard's campus is the Old Yard, which is a spacious, tree-lined lawn or quadrangle around which several of the oldest university buildings are located. However, the grass lawn and trees extend east behind the Old Yard and are surrounded by freshmen dormitories, Widener Library, and the Memorial Church and are known generally as Harvard Yard. These spaces are central to the visual impact and the enduring memory of Harvard University. Like the interior courtyards and open parks of Oxford and Cambridge, they provide places for formal gatherings, extramural athletic competition, and simple restful contemplation. They also are the environment for the structures and provide a colorful visual backdrop, which complements the primarily red brick buildings. The visitor to Harvard, the student, and the faculty member all must feel powerfully influenced by these visually important details.

The buildings in the area of Harvard Yard are examples of the Georgian or Federal Colonial style. Derived from Palladio by Inigo Jones in England, as mentioned in Chapter Two, this is a comparatively unadorned type of architecture which emphasizes symmetry, uninterrupted horizontal and vertical lines, and solidity. Massachusetts Hall is the oldest building still standing on the campus and includes both administration and dormitory space. It is a simple four story rectangle, whose top story is angled and slated with dormer windows. Paired chimneys stand at each end of the structure as well as in the middle. The Faculty Club is similar in appearance but has only two stories and a single chimney. Widener Library departs from the Federal Colonial style with Neoclassical architecture. Large and imposing, the entryway features twelve Corinthian columns and comparatively steep stone stairs. Facing the library, the much smaller Memorial Church has four Doric columns under a plain Greek pediment and a steeple in the manner of the Church of St. Martin in the Fields in London. Emerson Hall is another rather massive building in the Neoclassical manner, which is three stories tall. It features six Ionic engaged columns on its entryway and utilizes the typical red brick even for the columns. Although not all the buildings in the central area of Harvard University are the same architectural style, they share powerful visual components—color, symmetry, strong vertical and horizontal lines, and solidity.

Fig. 7-15 Massachusetts Hall

Outside the central area of the Harvard campus are numerous types of buildings, including churches, research centers and laboratories, art museums, and design centers. Some of these buildings represent Modern or 20th century architecture at its best. Carpenter Center for the Visual Arts, the only building in the US designed by Le Corbusier, is perhaps the best example to discuss. It is a large off-white stone building with a generous mixture of straight lines, curves, and glass. As a center for the visual arts, this building certainly positively reflects its domain, and it manages to avoid some of the perils of the Modern style emphasis on form or function alone. The Arthur M. Sackler Museum down the street is a rather small building which is essentially rectangular and features a brick surface. The color of the brick is not the typical red of the central campus but is instead an orange tone. The Sackler Museum building lacks some of the grace of the Carpenter Center in that it is almost devoid of anything but straight lines. Its entryway, however, does feature visually attractive contrasting white stone work, a glass and steel entry portico, and two flanking white stone pillars.

Fig. 7-16 Carpenter Center for the Visual Arts

Texas A&M University and Sam Houston State University

Harvard is but one of hundreds of private and public universities in the US, most of which would be worthy of assessment as compelling architecture. It is obviously not feasible to assess them all, therefore two universities with which the author is especially familiar, Texas A&M University in his home town and Sam Houston State University where he is employed, will be used to illustrate two important aspects of architectural elements used for their power to influence. Neither university, unfortunately, has managed to follow a uniform style of architecture nor always achieve the Vitruvian ideal of beauty. Like many elementary and secondary schools in America, not to say many industrial and commercial enterprises, A&M and SHSU have frequently built only for the sake of strength and function. Both universities, however, show evidence in several instances of an understanding of the importance of beauty.

One of the most majestic features of the Texas A&M campus, for example, is the Administration Building and its impressive approach. This very large university has a sprawling campus that was originally located on a hill near the city of Bryan, Texas. According to tradition, the parents of A&M students wanted their offspring to concentrate on their education and not to succumb to the "sinful" temptations of the bars and music halls of Bryan. Texas A&M, therefore, was built far enough away from Bryan to make a trip there difficult on foot. A railroad stop—College Station—was built near the campus and eventually a city by the same name grew up around the university. Today, Bryan and College Station have grown to meet each other as twin cities.

Fig. 7-17 Texas A&M Approach

Not only is Texas A&M a very large university, enrolling nearly 50,000 students, its campus features many high rise buildings and others of considerable mass. The impression overall is one of size and power. The cornerstone of this impression, however, is undoubtedly the J. K. Williams Administration Building and its street approach along New Main Drive. Many important structures gain architectural power by their location on a height. The Acropolis in Athens is an example. Others gain power by their location along or at the end of a prominent avenue. Seen from above, the avenue appears as a strong horizontal line or arrow pointing to the structure. The Louvre, located at one end of the Champs Élysées, is an example, as is the Prado in Madrid. New Main Drive in College Station leads down a gentle slope at a 90° angle from a primary through street to the sprawling Texas A&M campus, passing through open green areas on either side. On the left is a golf course, and on the right is a former parade ground. The visual message is clearly that one has left the world of the city and entered one of the groves of academe.

The building itself is large and imposing as if representing the authority and strength of the university. It is a four story Neoclassical building with strong horizontal lines. It is also perfectly symmetrical. In addition to being at the end of New Main Drive, the Williams Administration Building is built on a small knoll fronted by a regularly maintained flower garden. Stone stairs ascend to the entrance on the second floor the façade of which is formed by 14 majestic Ionic columns. The stone work of the first floor is rugged, while that above is smooth. All of it is off-white. At night, the majesty of this building is accentuated with portico lights and flood lights. Students, faculty, and alumni of Texas A&M may have differing choices for their favorite campus building, but from the standpoint of visual persuasion the J. K. Williams Administration is outstanding.

Before there was a Sam Houston State University, the hilltop location of the university was occupied by Austin College until the school moved to Sherman, Texas (http://www.shsu.edu). The principal building of the college, today known as Austin Hall, has become the central architectural symbol of the university and is the oldest building used continuously for educational purposes west of the Mississippi River. Its cornerstone was laid in 1851 (http://www.buildingshsu.com). Today a refurbished Austin Hall is used for receptions, formal meetings, and other special occasions. It stands at one end of the university's central quadrangle and is fronted by a bronze statue of Sam Houston, first President of the Republic of Texas and later both Governor of and Senator from the state of Texas.

Austin Hall is the most visually impressive building on the SHSU campus. It is a two story structure done in the Federal or Georgian Colonial style with a Neoclassical entry portico of three Doric columns facing south. The building is a rectangle, and there is a matching portico facing north. The stonework is red brick, and the window trim, columns, and roof cupola accent the brickwork with white. Visually, the building is symmetrical and imposingly massive, acting as a sort of anchor for the campus. Over the years, Austin Hall was considered for destruction, but perhaps for its visual power it has endured. An adjacent Gothic style building known as Old Main burned to the ground and Austin Hall was damaged. However, it was repaired so that it shows no evidence of the fire.

Fig. 7-18 Austin Hall

One of the earliest buildings of the institution that was to become SHSU was the Peabody Memorial Library built in 1902 to house the library collection which had outgrown its space in the administration building known as Old Main. A newer library has now supplanted this one, so it is now also used for special events and university archives. The library building sits directly across from Austin Hall although a bit downhill. It has a ground floor and a basement, and its floor plan is a Greek cross. Built of the same red brick as Austin Hall, the library's distinguishing features architecturally are its succession of arched windows all around the building and the stained glass which remains at the top of several of them. Sam Houston State University's more recent architecture has been quite eclectic, with no other buildings which are quite the visual equals of Austin Hall and Peabody Memorial Library. In the final section of this chapter, universities across the Pacific Ocean in Australia and New Zealand will be examined for their architectural power.

Universities in Australia and New Zealand

Melbourne, Australia, lies on the coast southwest of the country's largest city, Sydney, and is home to two important academic institutions. These are the University of Melbourne, with a large central campus north of the city's central business district, and RMIT University, the Royal Melbourne Institute of Technology. RMIT also lies north of the Melbourne central business district but south of the main campus of the University of Melbourne. Mclbourne University's campus is similar to that of Harvard, Texas A&M, and many others that are separated out as single spaces. Its architecture is traditional with many buildings patterned after Gothic ones in England. While these buildings are surely architecturally interesting, they are no more so than those already considered in this chapter. The exception is the Ian Potter Center, the very modern building downtown in Federation Square which houses the university's collection of the visual arts and which was examined in Chapter Three.

Fig. 7-19 RMIT

RMIT University also has a number of buildings but lacks anything like a yard or quadrangle. Instead, RMIT gives the impression of blending in with the buildings all around it. This does not mean that it is not visually striking however. One of its buildings, called simply Building 8 is the most visually striking one and is the home of the university's student unions. It is fortunate indeed that it is possible to provide an image for this building, for it nearly defies verbal description. Additionally, it is very colorful, a quality which if of course lost in a black and white image. Its façade is light colored stone and dark glass over a several story structure interspersed with geometrically diverse window projections. Also interspersed along the upper façade are alternating pink and aqua geometric figures. The entry portal features rounded angular forms, but the overall strength of the building comes from its strong horizontal lines. Perhaps this building would not be thought of as visually persuasive to every viewer. It is unusual and perhaps quirky or whimsical. However, the structure easily commands the viewer's attention and is clearly appropriate to an institute of technology and to a set of organizations devoted to sports and other student activities in a urban setting.

"Across the ditch" on New Zealand's North Island, Victoria University of Wellington presents a campus reminiscent in many ways of the University of Melbourne. It has a main campus located in Kelburn, which is a hilltop suburb of Wellington, but there are other locations of university buildings including Rutherford House in the city proper adjacent to the Wellington Railway Station and near the Parliament district overlooking Lambton Harbor. While the main campus location enjoys the prominence of a hilltop site, Rutherford House anchors the Pipitea campus in an equally prime spot. It is readily accessible to the Wellington central business district, it offers a compelling view of the town's harbor, and it is adjacent to the power of the nation's government.

Fig. 7-20 Rutherford House, Victoria University

As in Melbourne, the main campus buildings at Kelburn are for the most part traditional architecturally, but Rutherford House is strikingly modern in its use of steel and glass, strong vertical lines, color, and a smooth metal curved entryway. Although not a skyscraper, it gives the visual impression of a commercial center, thus making it fit well with the nearby business district. At the top of Rutherford House are large red panels interspersed with a dark gray one on which the name of the university appears. While the Kelburn campus should give its students and faculty an atmosphere of traditional learning through its architecture, the Pipitea campus captures the stylistic nuances of contemporary life and work.

Universities are no less powerful as compelling visual forms than any of the other structures examined in this book. The emphasis in this chapter has been on individual buildings, but it is clearly the whole campus of a university that visually influences its faculty, students, staff, and alumni. Additionally, universities are iconic for the cities and larger areas in which they are located. They lend prestige in the same manner as museums and public buildings in general. With this in mind, the subject of the next chapter is those very public, iconic buildings.

CHAPTER EIGHT

CITADELS OF POWER: NATIONAL, STATE, REGIONAL, AND LOCAL GOVERNMENT BUILDINGS

"There is no greater expression of a civilization's fundamental values and highest aspirations," observes Andrews (2006, 20), "than its public architecture." Although his subject is the courthouses of Texas, Andrews might as well have been speaking for public architecture at every level. In other places as well as Texas, "magnificent governmental buildings reflect the independence and fortitude of a resolute people determined to create order and permanence out of a vast wilderness." Andrews points out that "the courthouse square became the focal point for celebrations, markets, parades, and innumerable civic and public functions. Memorials to a county's war heroes and most sacred causes were erected on the courthouse grounds. The courthouse became an expression of a community's character and a reflection of its worth and confidence." These observations about Texas courthouses are often true for the other government buildings to be considered here.

National, state, regional, and local government buildings exist for a variety of functions. At the national or federal level, they serve as meeting places for legislators and judges, as well as for trials, and the same is true at the state and local levels. At the regional level in the US, courthouses include offices for county government officers, such as tax collectors and clerks, jury rooms, and document storage areas. Among the documents which both originate and are stored in courthouses, there are deeds of ownership, birth, marriage, and death records, regional archives, and so on. It is not at all difficult to understand the primary persuasive message of government buildings. Simply put, it is power.

From the earliest fortresses of humankind to the most recent parliament building, the idea has been to suggest visually that this building is the seat of the greatest power in the land or the region or the community. There are also several other related persuasive messages. Another, for example, is authority. Government buildings stand for the rule of law. Respect is part of the equation as well, since the law must be acknowledged in order to be effective. The buildings also are about control. As repositories, their essential documents are necessary for the orderly conduct of the community's affairs. And finally perhaps, a government building should convey visually the status of the entity it represents. In this way, the architecture of government is culturally iconic, much as great museums are.

To exude power and authority, to command respect, to exert control, and to stand as cultural icons, governmental buildings must be visually commanding. They must be located in prominent sites, be massive and grand, be imposing, demonstrate strong horizontal and vertical lines, and utilize other of the most powerful elements of visual form. In this chapter, the buildings to be examined include houses of parliament, national government buildings, state capitols, county courthouses, and city halls. While as in previous chapters it is not feasible to be comprehensive in coverage, it is possible to be suggestive with a variety of examples.

Beginning at the national or federal level, some of the most visually compelling houses of parliament and chambers of deputies in the United Kingdom and Continental Europe will be considered. These include the Houses of Parliament in London and Edinburgh, the Reichstag in Berlin, the Congress of Deputies in Madrid, and the Parliament buildings in Budapest and Vienna. Then, the US Capitol and the US Supreme Court buildings will be considered, followed by New Zealand's national government complex in Wellington. State government buildings will be the next consideration and will include Parliament House for the Australian state of Victoria in Melbourne, the State Library of Victoria, Ali'iōlani Hale, housing the Supreme Court of Hawai'i, and the State Capitol of Texas. In the US, an especially important group of buildings are the courthouses representing the seat of power for the county subdivisions of state government. Texas may well have the most visually distinctive set of such courthouses in the country if not the world. Of the many formidable courthouses built in Texas in the last half of the 19th century, the ones to be examined include those in the counties of Bell, Hill, Ellis, Tarrant, Dallas, and Denton. Town halls will follow and will include Brussels's Town Hall and the Town Hall of Vilnius in Lithuania. The beginning point for this chapter is one of the most iconic and recognizable building complexes in the world: the British Houses of Parliament in London.

National Government Buildings of the United Kingdom

The British government is a parliamentary monarchy divided into two legislative assemblies: the House of Lords and the House of Commons. Together, these two bodies meet in the complex of connected buildings known collectively as the Palace of Westminster, which also includes the clock tower known popularly as Big Ben. There are several prime sites for public buildings in London, but surely there is not one more auspicious than the location of the Palace of Westminster. It sits directly on the Thames River at the foot of Whitehall, which leads past several other imposing government buildings, past Downing Street and the Prime Minister's residence at No. 10, up to Trafalgar Square with its Nelson Memorial, art museums, and church. Across a street is Westminster Abbey. The building complex itself is regal, as is befitting what was intended to be a royal palace.

Fig. 8-1 Palace of Westminster

The present complex forms a large rectangle, much like the arrangement of buildings within castle walls. The location has been in use as a site for royalty and government since the 11th century. Fires and the air raids by Germany in World War II have resulted in much alteration, expansion, and rebuilding. The present complex, however, is very consistent architecturally and follows the Perpendicular Gothic style that was examined in the case of King's College Chapel at Oxford in Chapter Six. It is visually striking enough to have been the subject of a painting by Claude Monet.

There are three main towers dominating Westminster. At the southwestern corner of the complex is Victoria Tower, which is the largest and at the base of which is the Sovereign's Entrance to the palace. This is a square tower with turrets at each corner and two stories of arched windows. The central tower, the shortest of the three, is steeple-like and adorned with pinnacles. Third, at the north end of the complex is the clock tower one of whose five bells is known as Big Ben. Apart from the palace itself, Big Ben has become an iconic structure itself. Architecturally similar to the campanile of St. Mark's in Venice, Big Ben has been used on post cards, as the model for tins of cookies and candy, and the emblem of many British enterprises. Shorter square towers adorn the two corners on the Thames side of the palace. Westminster shares many of the familiar characteristics of the Gothic architectural style. Especially noteworthy are the pinnacles regularly spaced along the sides and the regularly repeated array of rectangular windows. The stonework is a shade of tan, which is particularly reflective in both the morning and evening sun.

Scotland has been a part of Great Britain since the Acts of Union joined it with Wales and England in 1707 and a part of the United Kingdom, which also includes Northern Ireland, since 1800. As such, Scotland's representatives meet in and its laws are made at the Houses of Parliament in London (http://www.wikipedia.org). Each member of the United Kingdom, however, has a devolved unicameral legislature wherein matters not reserved to the United Kingdom at large can be considered. This arrangement is not unlike the powers reserved to the states by the US government. There is at this date a growing interest in Scotland in independence, and at least partly for that reason the Scottish legislators decided to build the Scottish Parliament building in Edinburgh which opened in 2004.

Edinburgh's Royal Mile, extending from Edinburgh Castle at the high end to the Palace of Holyroodhouse at the low end, has been discussed in previous chapters with respect to the Cathedral of St. Giles and the University of Edinburgh. In the following chapter, both the castle and the palace will be considered, but clearly this is the most significant avenue in Scotland for many reasons. Also at the low end of the Royal Mile and essentially across the street from the grounds of the Palace of Holyroodhouse is the location of the new Parliament Building. As befits

its recency, it is a thoroughly modern structure of oak, steel, and stone which beggars verbal description.

The Parliament Building was conceived by the architect, Enric Miralles, as growing out of the land, so that many of the forms in the building which do not reflect traditional views of architectural elements instead reflect natural items such as leaves. Viewed from above, for example, the shape of the roofs of the component structures resemble leaves (http://www.designbuild-network.com). The public entrance to the building utilizes curved as well as angular overhangs, an awning of sorts made of tubing and wood, and wooden slats covering various doors. To the right of the entrance and along one street side of the building, the upper two stories of the structure jut out and seem to hang in space. Along the ground floor, curved stone is adorned with depictions of street scenes and written quotations. A recurring theme is expressed in the cladding along the upper stories in the form of firearms triggers.

Fig. 8-2 Scottish Parliament

The Scottish Parliament Building is unique in its appearance and lacks many of the architectural elements discussed in Chapter One which contribute to compelling form. Its place in the history of architectural styles is occupied by only a few other modern structures, notably the Guggenheim Bilbao Museum and the Walt Disney Concert Hall in Los Angeles, both of which were designed by Frank Gehry. The Parliament Building certainly is not, however, either ugly or lacking in visual power. Of course, it is located on an auspicious site as noted above. Second, it uses building materials that very much emphasize the land, especially through the interplay of wood and stone. Although the curves, angles, and asymmetrical lines sometimes appear to clash, the overall visual effect is appealing. The basic off white, gray, and brown colors complement the overall solidity of the building. The building does not in any way fit its surroundings, which are virtually all traditional pieces of architecture. However, the very unusual nature of this building made it instantly iconic and focused attention on the Scottish parliament in a powerful way. By contrast, Continental European government buildings are traditionally powerful visually.

National Government Buildings of Continental Europe

Of all the national government buildings in Europe, the German Reichstag in Berlin may be the best known. This is not so much for the architecture of the building, however, as for the building's checkered past. After the establishment of the German Empire in 1871, the need for a building in which the Parliament could meet became paramount, and the Reichstag was completed in 1894. Arsonists burned the building in 1933, and Allied bombing of Berlin in World War II along with Russian artillery left it a ruin. During the Cold War, some renovation of the building was done just to keep the building from collapsing completely. Although the Reichstag was located in West Berlin at this time, complete renovation came about only after the collapse of Communism and the reunification of Germany (http://www.aviewoncities.com/berlin). Today, a fully restored Reichstag is on view and is the meeting place for the Bundestag or parliament of Germany.

The Reichstag is located centrally in the reunited city of Berlin, sitting in the shadow of the Brandenburg Gate. This location formerly marked the line between West and East Berlin during the Cold War. Behind the Gate lies Unter den Linden, Berlin's version of the Champs Élysées in Paris, the fine Hotel Adlon, and many of Germany's best museums. To the west, this street, called Strasse des 17 Juni, stretches on through the Tiergarten and circles around the Siegessäule or Victory Column. This location, then, is ideal for the architectural implication of power. The building itself is a rather traditional Neoclassical structure. The elevated entrance is reached by a set of stairs and features six Corinthian columns mounted by a typical Greek pediment and sculpted tympanum. Along the entablature are inscribed the words "Dem Deutschen Volke" ("To the German People"). On each of the four corners of the rectangular structure are towers and in the center a glass dome now replaces the original enclosed one heavily damaged during the war. Corinthian engaged columns continue along the exterior walls on either side of the entrance. The Reichstag has powerful horizontal lines and an attendant mass and solidity. Even without the historical events which made it famous, the structure is a powerful visual symbol of governmental influence. South of Berlin in Vienna is another similar parliamentary building with similar power.

Fig. 8-3 Reichstag/Bundestag

The Austrian Parliament meets in Vienna in a building that fits in well with the multiple other monumental buildings of the city. The building is massive, and its site near the Hofburg Palace and other government buildings and museums is auspicious. The structure is certainly the architectural and visual equal of its surroundings and is more ornate than most of the other buildings nearby. It is Neoclassical in style, its entryway in fact being a copy of the entrance to the Erechtheion on the Acropolis in Athens (http://www.gothereguide.com). This main entrance is inside a portico with eight marble Corinthian columns whose capitals are decorated with gold leaf. A semi-circular drive leads up to the entrance, which is on the second floor. Above the colonnade is a pediment whose tympanum is

Fig. 8-4 Austrian Parliament

decorated with the sculpture of figures from the period of the Austro-Hungarian Empire. A fifteen foot sculpture of the goddess Athena sits in front of the Parliament Building and is also adorned with gold leaf. To the sides of the entry portico, the rectangular building extends horizontally and is faced by Corinthian engaged columns alternated with windows. The effect is particularly rhythmic and stately.

In nearby Budapest, Hungary, there is another powerful example of the visual persuasiveness of architecture, although it is Gothic in style rather than Neoclassical. This is the Hungarian Parliament Building. The capital of Hungary was created by joining together Buda, the hilltop area overlooking the Danube River, and Pest, the flat area across the river. On the Buda side sits the Royal Palace, which will be examined in the next chapter, and opposite it is the Parliament Building sitting directly on the river. Several architecturally significant bridges link the two parts of the city, the Chain Bridge being the closest to the Parliament building. This site is a powerful element in the visual impressiveness of the building, both in terms of other nearby buildings of prominence and the presence of the river. The effect is quite like that of the Palace of Westminster sitting on the Thames River and adjacent to significant bridges. Indeed, the Hungarian Parliament Building resembles Westminster to a limited degree.

Fig. 8-5 Hungarian Parliament

The Budapest structure is, like Westminster, a horizontally spread out complex of connected buildings. It has the usual pinnacles of Gothic buildings, but it features a central dome which is interestingly framed by flying buttresses. Shorter tower like structures, with spires at each corner, flank the dome. The stonework is largely off white, but there is also a liberal use of maroon trim and roofing. The main entrance opposite the river side is not merely a portico but a wing of the building which projects outward and features stairs leading upward between two sculpted lions. The river side of the building and the one most frequently photographed is a façade of repeated curved archways and slender columns much in the manner of the Ospedale degli Innocenti in Florence, Italy. This is, of course, Italian Renaissance architectural style, not Gothic. The center of the façade, however, does have two Gothic spires which frame the central dome.

The final Continental European government building to be noted here is the Congress of Deputies in Madrid, Spain. Although a national government building, the Congress of Deputies houses only the lower chamber of Spain's legislative body. The Senate meets in a separate building closer to the Royal Palace. The Congress of Deputies is located just to the east of the center of Madrid on a hill above the Prado Museum. As such, the site of the building is commanding, even though it is crowded around by other buildings. The building itself has great mass and solidity, which is underscored by heavy Neoclassical features: fluted Corinthian columns at the top of a stone staircase, a Greek pediment with a tympanum of carved figures, and solid, smooth, off white exterior walls interspersed with simple windows. Save for the crowding imposed on it by surrounding buildings, the Congress of Deputies in Madrid is quite compelling. Of all the government buildings being considered here, and perhaps of all such buildings in the world, the only one which approaches the Palace of Westminster in iconicity is the United States Capitol in Washington, DC.

Fig. 8-6 Congress of Deputies

National Government Buildings of The United States

As with the Palace of Westminster, the r is the meeting place of both the Senate and the House of Representatives in the bicameral legislature. Collectively, the two are most commonly referred to as simply Congress. The building evolved over a number of years and, like Westminster, has been damaged by cannon fire, been burned and rebuilt, and has also been expanded. The architecture is nonetheless remarkably consistent. The building alone is magnificent, but its location is as architecturally compelling as the structure itself. It stands at the east end of the National Mall, a great, open, park like field bounded on each side by the museums of the Smithsonian Institution and facing the Washington Monument on the west end of the mall. Beyond the Washington Monument on a straight line is the US World War II Memorial and the Lincoln Memorial with its Reflecting Pool. Although the visual impressiveness of this layout can only be completely appreciated from above, its sight lines permit any visitor to take in the significance of the radiating lines and surrounding structures. Seen from the Washington Monument, the US Capitol rises to its small hill location and stands in radiant white majesty at the end of an unobstructed vista.

Fig. 8-7 US Capitol

Like Westminster, the US Capitol is massive and has strong horizontal lines suggesting strength. It is also absolutely symmetrical, whether one is facing the East or the West Entrance. The architecture is Neoclassical, although the huge and widely visible dome is suggestive of Michelangelo's dome on St. Peter's Basilica and Christopher Wren's dome on St. Paul's Cathedral in London. Centrally, then, the Capitol also has a strong ascending or vertical line. At both the north and south ends of the building are matching wings, which on the east front have stairs leading to a columned entry portico topped with a Greek pediment. On the west front, these wings have arched windows on the ground level and columns all along the stories above.

The central entrance on the east front of the Capitol mimics the entry porticoes on the wings. On the west front, there are two stairs leading from the mall to the building itself divided by lower stories of the Capitol and topped by a colonnaded entrance between the stairs. Rising above the center of the building is the Capitol's most powerful visual symbol: the huge cast iron dome. At its base, the dome is ringed with columns. The next level is a ring of arched windows, then the dome curves upward to a colonnaded cupola topped with a statue of Freedom.

Located nearby are a number of other architecturally impressive government buildings, such as those of the Library of Congress and numerous buildings housing specialized departments of the government such as Agriculture. An especially compelling one is the US Supreme Court building. Also Neoclassical in design, the structure's entry portico at least resembles nothing more than it does a Greek or Roman temple, such as the Maison Carrée in Nîmes, France. The building dates from 1935 and was designed to complement the surrounding architecture. Marble stairs lead one up to this temple of justice, being flanked on each side by walls and two statues. The left statue is called the Contemplation of Justice, while the one on the right is called the Guardian of the Law. The portico has sixteen marble Corinthian columns and a pediment decorated with a frieze of sculptures of Liberty, Order, and Authority. The east or back side of the building has a similar but less ornamented entry dominated by four Corinthian columns flanked by two Corinthian engaged columns on each side. The sculptures of the eastern pediment are of Moses, Confucius, and Solon in the center flanked by figures representing such philosophical ideals as the "Means of Enforcing the Law" (http://architecture.about.com). The four story building spreads out horizontally on each side of the entryways and is

Fig. 8-8 US Supreme Court

largely the same off white Vermont marble of the columns and stairs. The power of the law is represented well by the strong horizontal lines of the Supreme Court Building and the temple-like nature of the architecture. The last task before turning attention to state, regional, and local government buildings is to examine the national government complex in Wellington, New Zealand. Located on the southern tip of the North Island, Wellington is cradled in a bay at the foot of the mountain range which dominates the country. One might have mixed expectations about the nature of its three principal national government buildings. On the one hand, one might expect to find radically new forms of architecture befitting a comparatively recently settled land, and of course this is the case with Wellington's museum Te Papa Tongarewa as noted in Chapter Three. On the other hand, the expectation might be that the government buildings would mimic those in Great Britain, from which most of the new settlers came. What one actually finds is both, with the cause probably lying in the era in which the respective buildings were designed and built. The one with the most traditional architecture is Parliament House itself.

Fig. 8-9 Wellington Parliament and Beehive

National Government Buildings of New Zealand

Parliament House is a traditional three story Neoclassical building with an entryway framed by four Doric columns on the second floor at the head of a stone stairway. Doric columns continue along the exterior walls to the right interspersed rhythmically with slender rectangular windows. There is no pediment. The first floor stonework is rough, while the upper stories are smooth, and the visual effect is of great solidity. In place of a matching side of the building to the left is a much newer wing done in Modern style. This building or wing is known popularly as "the Beehive," and one look explains this sobriquet. Vertical supports of different sizes dominate each floor and continue around the circular structure giving it a strong vertical presence. In spite of its modernity, the building still conveys a sense of solidity and size.

Fig. 8-10 Wellington Parliamentary Library

North, or to the right, of Parliament House and the Beehive and at the very top of the hill on which these government buildings sit is the most eye-catching of the three. The Parliamentary Library is done in a subdued Gothic style and painted exterior features. It is white masonry decorated with mauve accents. A single stairway leads up to the two story building, which has an arched portico for the entrance topped with a triangular peak. Two more triangular peaks lie on either side, and the roofline is dotted with small Gothic pinnacles. While the Library may not convey quite the solidity and mass of the other buildings nearby, it certainly adds the visual impact of height and color. All three buildings lie at the top of a hill above Lambton Harbor and dominate the upper central city skyline.

State Government Buildings

Where there are countries with political divisions called provinces or states, there are government buildings analogous to the national government buildings discussed previously. To account properly for the visual persuasiveness of these buildings is a task well beyond the scope of this work. In the US alone, there are 50 such buildings. However, it is fair to say that the types of visual persuasion represented by state government buildings, as well as regional and local ones, do not vary wildly from the national ones already considered. This section of the chapter, therefore, will consider a typical Australian state capitol, the Parliament House of Victoria in Melbourne, the State Library of Victoria, and two of the most distinctive state government buildings in the US, the Hawai'i Supreme Court Building and the Texas State Capitol.

Fig. 8-11 Parliament House of Victoria

Melbourne, Australia, the capital of the state of Victoria, is a city of hills, with much of the city's central business district, as noted in Chapter Three, sitting beside the Yarra River at the foot of one of these hills. Above the city center at thc top of Bourke Street is the Parliament House of Victoria. A massive structure with strong horizontal lines, the building dominates the hill on which it sits. Broad stairs lead to a porch defined by nine fluted Doric columns. Behind these, there are more columns and pilasters flanking the doorway. A decorated entablature is situated above the columns, but there is no pediment. More columns range along the front of the building, and at each end a wing of the building projects forward. Decorative capitals adorn the roofline and the entablature. There is no dome, although one was originally projected. However, the site and mass of Parliament House is powerful architecture.

Fig. 8-12 State Library of Victoria

The State Library of Victoria, if anything, is a more architecturally impressive building than Parliament House. This structure is located uphill from the central business district along Swanston Street. The building is a perfectly symmetrical Neoclassical structure with wings at each end. Broad stairs lead to a Corinthian colonnaded entrance. Eight columns front the entry portico and more range along the front of the building on either side. The temple-like building features a typical Greek pediment. Two main stories are topped by a roof bordered with a banister. An octagonal dome rises three stories above the main building providing an enormous reading room within. In many ways, a library is a museum, and this building conveys the sense of elegance and stateliness that is associated with the idea of a "temple of the muses."

Hawai'i is unique among the states of the United States of America. It is a group of islands in the Pacific Ocean southwest of the US mainland, which was once ruled by royalty. Numerous distinctive buildings from this royal period remain in use today for the purposes of government. One of these is Ali'iōlani Hale, housing the state supreme court, originally intended as a royal palace for King Kamehameha V whose statue is located in front of the building. Its architectural style is Italian Renaissance Revival and features an abundance of Ionic columns. The building is symmetrical with a central entryway and square tower rising four stories into the air. Each of the first three stories features paired columns, and the fourth story is a clock tower. Beside this central tower are porches ranging out to each side and fronted with more Ionic columns. At each end of the building are projecting wings of plain stone. A balustrade lines the roof. The architecture is compelling, to be sure, but the style is more European than Hawaiian. In the following chapter, however, a more indigenous architecture, that of the 'Iolani Palace, will be examined.

Fig. 8-13 Ali 'iōlani Hale

The state of Texas is also unique in many ways. It has the distinction of being the only part of the US that was a country. For a decade following the Battle of San Jacinto and independence from Mexico in 1836, the Republic of Texas had embassies in Washington, DC, London, and

Paris. Until the addition of Alaska to the Union, Texas was the largest of the states. It shares a large part of its essential character with other western states such as New Mexico and Arizona by having vast areas that are arid, mountainous, and desolate. The early settlers of Texas had to have been hardy people to endure privation, threats of marauding Comanches, and an inhospitable climate, not to say the continuous threat, for many of Texas's early years, of invasion by Mexico. Because of these things, governmental buildings in Texas were perhaps even more important in the minds of its citizens than they were in other states. As a result, it seems also to have been more important than in other places to construct buildings of compelling architectural form.

Fig. 8-14 Texas State Capitol

The State Capitol Building in Austin is a massive and architecturally grand structure nearly rivaling the US Capitol compared to which it is second only in size. It is the largest of the state capitols in the US. Although not situated at the highest point in Austin, the capitol is located on a gentle rise at the end of what is perhaps the main street of the city: Congress Avenue. All around the building there is green, open, park-like space complementing it. A broad walkway leads from the end of Congress Avenue to the entrance, which faces south. An elliptical drive parallels the walkway on either side. The architecture resembles the US Capitol, having been done in Renaissance Revival style (http://www.tspb.state.tx.us/SPB). Erected in 1888, the foundation of the capitol is limestone, while the building itself is red granite. The rectangular structure is perfectly symmetrical with western and eastern wings as well as the central portico projecting out from the main floor plan. The entryway has a single curved arch topped by a Greek pediment and flanked by four double pilasters. Rising over the entryway is the central dome, virtually a replica of the US Capitol dome although 15 feet taller. The north side of the building is similar to the south, except that it has a colonnaded portico with three arches rather than the single one of the south entrance. The grand structure that is the Texas State Capitol is by no means the only especially compelling such government building in the state.

Courthouses of Texas

Texas has 254 counties, with perhaps 200 having courthouses worth noting architecturally. Moe (2006, 16) writes that "the Lone Star State's collection of these landmarks is absolutely unmatched in the diversity and quality of their design." "Many of them," he continues, "are visual knockouts, boasting enough towers, turrets, domes and porticoes to satisfy the most jaded building aficionado." Texas courthouses, like those in other states, "were intended to be brick-and-stone embodiments of the stability of democratic government and the awesome majesty of the Rule of Law. Architects worked hard to make them 'fitting,' and to ensure that visitors would feel uplifted by the mere act of stepping inside to conduct business in their hallowed halls." In this way, Texas courthouses are especially good examples of architecture as visual persuasion. Especially notable in this respect are the courthouses of Bell, Hill, Ellis, Tarrant, Denton, and Dallas counties. By virtue of their importance, all are located centrally in a city.

One of the most magnificent courthouses in Texas is in the city of Belton in Bell County. "This stately Renaissance Revival temple of justice more closely resembles a state capitol than a seat of local government. The building's most distinguished characteristic is its grand clock tower with a columned gallery, which rises 125 feet" (Andrews 2006, 40). The floor plan is square—courthouses are located on squares of land—and has entrances on each of its four sides. Each entrance has a projecting portico on the second and third floors of which are four Corinthian columns topped

Fig. 8-15 Bell County Courthouse

by a Greek pediment. The aforementioned clock tower has clocks on each of its four sides and is topped with a dome and a statue of Justice. The building is of rough limestone on the ground floor and smooth stone on the upper stories. The color of the building is off-white, which shines in the Texas sun and is set in contrast to the lush greenery of the surrounding lawn and trees.

Northeast of Belton is the city of Hillsboro featuring the Hill County Courthouse. The original courthouse was built in 1890 but was destroyed by fire in 1993. In evidence of the claims made earlier in this chapter about the importance of courthouses to a community, the citizens of Hillsboro accurately restored this grand old building and reopened it in 1999. The building is a square like the Bell County Courthouse and like it has entrances on each of its four sides. It is also three stories high, but its bell and clock tower rises another four stories over the center portion of the structure. Fluted and banded Corinthian columns decorate each entrance, and their slender and tall forms give the building a particularly strong vertical impression. Slender arched windows range within and along the sides of the entryways adding to this verticality. As in Bell County, the building uses rough off-white limestone. Architecturally, the "building is a Second Empire design with Italianate and Classical Revival influences" (Andrews 2006, 66).

Fig. 8-16 Hill County Courthouse

Continuing northeast from Hillsboro, one comes to the city of Waxahachie, the home of the Ellis County Courthouse. This courthouse has its entrances on each of the four corners rather than the sides. These entrances are arched and lead to columned porticoes. The Ellis County Courthouse is a three story building with a clock tower rising another six stories in the air. The clock towers which have been noted so far in this chapter not only give the courthouses a strong sense of verticality, but they allow the "citadel of power" to be seen at a great distance. In this respect, they resemble and function like the towers of the great Gothic cathedrals and the steeples of churches. The building is multicolored, having been built of "Burnet County limestone, Pecos red sandstone and rose pink granite" (Andrews 2006, 122). The courthouse is well-known in Texas tradition in that 21 carved stone faces are interspersed in the red sandstone around the building. This stone carving was commissioned to German artisans, and there are legends that some of the faces are of one of the men's lost loves. The architecture is largely Romanesque, and the courthouse lacks some of the grace of those in Bell and Hill counties. However, it loses nothing to these other courthouses in its imposing mass and solidity.

Fig. 8-17 Ellis County Courthouse

The remaining three courthouses to be considered here lie in a triangle north and northwest of Waxahachie, which is commonly known as the Metroplex. The two central cities are Dallas and Fort Worth, which are linked by interstate highways, but both between and north of Dallas and Fort Worth are multiple smaller cities and communities. Fort Worth is Texas's old "cow town," while Dallas is its larger and more urbane commercial center. North of Fort Worth and Dallas lies Denton, and each of these three cities is home to a visually compelling courthouse. Fort Worth is the location of the Tarrant County Courthouse, while Dallas has the Dallas County Courthouse and Denton the Denton County Courthouse.

Fig. 8-18 Tarrant County Courthouse

The Tarrant County Courthouse is located in downtown Fort Worth just at the northern edge of a comparatively new civic, cultural, entertainment, and shopping center known as Sundance Square. This area is filled with modern skyscrapers as well, which will be attended to in a subsequent chapter. The courthouse, of course, has been in place much longer than Sundance Square, dating to 1895, and the location is especially prominent architecturally. The site is a hilltop above the Trinity River at the center of downtown Fort Worth. As is often the case when there is rivalry between cities, "Tarrant County leaders were intent on building the biggest and best courthouse in the state, and certainly one more impressive than the new courthouse in neighboring Dallas County" (Andrews 2006, 112). Tastes vary, of course, but they seem to have succeeded. Unfortunately, "the final cost of the building so angered county citizens that every member of the commissioner's court was voted out of office in the next election."

The Renaissance Revival building which is the Tarrant County Courthouse is truly visually compelling. It is four stories tall topped by a 200 foot clock tower. The main entry portico features stairs leading up to four paired Doric columns. Above is a balustrade and six paired Corinthian columns topped by both arched windows and arched stonework. An ornately carved pediment sits at the roofline. The clock tower repeats the column and pediment theme and is finished by a dome. The rough stone used for the exterior is pink granite taken from the same location as the granite used in the building of the State Capitol discussed earlier in this chapter. The primary approach to the building is a gradual rise from Sundance Square, which emphasizes the building's vertical lines, but the courthouse also has strong horizontal lines that suggest solidity.

Fig. 8-19 Dallas County Courthouse

The Dallas County Courthouse, which the citizens of Tarrant County wished to outdo, is located just on the western edge of downtown Dallas. It is not located on a ridge, and the numerous skyscrapers which define the distinctive Dallas skyline rather outshine the building today. Additionally, the tall central clock tower of the original building was dismantled in 1919 for fear that winds would bring it down (Andrews 2006). While the present building lacks the visual impressiveness of the original, "Old Red" as it is called is nonetheless compelling. It has great mass and solidity accentuated by the use of rough red sandstone, red granite from Texas, and blue granite from Arkansas. The style is Romanesque, and each floor uses rhythmically repeated arched windows of varying size. Along each side are four turrets suggestive of English and European fortresses, including two of the original four griffins made of terra cotta.

Fig. 8-20 Denton County Courthouse

Last in this cursory study of Texas courthouses as types of visual persuasion is the Denton County Courthouse in the city of Denton northwest of Dallas and north of Fort Worth. Dating to 1896, the Denton County Courthouse is reminiscent of the Bell and Hill County courthouses in its use of off-white rough stone. There are Second Empire, Romanesque, and Italianate stylistic elements in the design as in the Hill County Courthouse. The clock tower is octagonal, and there are square towers at each corner of the roof. Each of the four sides has an arched entrance portico flanked by paired smooth granite columns, a theme repeated on each story above. The treatment of the examples of architecture in this chapter has proceeded down from the national level to the regional or county level, so it is at last the moment to consider local government buildings such as the town halls of Brussels, Vilnius, and London.

British and Continental European Town Halls

Fig. 8-21 Brussels Town Hall

Traditional architectural forms characterize the town halls of Continental Europe as well as elsewhere. As was the case with the Scottish Parliament Building, however, town halls built in the 20th century or later are another matter, which turns out to be the case with London's unusual one. In Brussels, Belgium, the Town Hall sits on the central square known as the Grand Place in the lower and older part of the city but features a central spire which can be seen from almost anywhere in Brussels. This building combines architectural elements from the Gothic style and the Italian Renaissance. Its ground floor has a delicate colonnaded portico with rhythmic rounded arches, while its two upper stories repeat the arch pattern in the windows. A central entryway has a single arched opening and is topped by a spire which rises another seven stories. There are Gothic pinnacles at each corner of the building and at each corner of the base of the clock tower. The effect is majestic and imposing. The Grand Place is ringed with other examples of architecture, but none dominates this central area more than the Town Hall.

Vilnius, the capital of Lithuania, is not a familiar location, as noted in Chapter Four. However, it boasts a Town Hall that is every bit as visually compelling as those in other cities in Europe. It is a spare Neoclassical design, sitting at one end of a stone paved square in Vilnius's old town. The building is a two story, off-white stone structure dominated by a single colonnaded entrance. A wide, shallow stairway leads up to a portico with six Doric columns with ornate capitals. There is also a typical Greek entablature and pediment. The tympanum features a single sculpted shield. It is probably the location and the clean elegance of the Town Hall that contribute most to its visual persuasiveness. Across the English Channel, London's Town Hall has a visual power which is uniquely modern.

Fig. 8-22 Vilnius Town Hall

The present London Town Hall was built quite recently and reflects some of the most innovative current architectural thinking. While its comparatively simple lines can be described verbally, one has the feeling that words alone would be insufficient to allow the listener to visualize this unusual building. It has had many detracters, as have a number of so-called Modern pieces of architecture. However, it has an undeniably compelling visual power. This is partly attributable to its location near the

Thames River in the south bank area called Southwark. It is readily visible from such other prominent sites as the Tower and the Tower Bridge.

Fig. 8-23 London Town Hall

The building can be described as appearing to be a football—not a soccer ball—tilted at an angle as if ready to be kicked. If thought of this way, it can further be said that the side of the football facing upward is smooth, while the downward tilted side is stepped. This side looks as if someone sliced the football into sections like a piece of fruit and stacked them one slightly forward of the one below. Each section corresponds to a floor of the multi-story building. The exterior is all glass and metal, and the interior features a spiral staircase complementing the exterior shape. There will continue to be debate about the attractiveness, indeed the appropriateness of this building, and it remains to be seen if this unique architecture will come to be widely copied. As will be seen in Chapter Elevan, there is a similar piece of architecture in London known colloquially as "The Pickle" or "The Gherkin," which resembles the Town Hall in its glass and steel construction and ovoid shape.

Whatever the *nachleben* might be for such structures as London's City Hall, there can be little question of the visual impact. This is unlikely to be the result of mere curiosity either. The shape is elementally a combination of curves, such as have been visually compelling throughout the history of architecture and are very basic to visual literacy. In spite of its apparent tilt, the Town Hall is solidly imposing and presents a vision of power albeit conceived in a most modern way. Power, authority, control, and iconicity are the hallmarks of government buildings across the globe and range from national to local institutions. The living and governing spaces of kings and queens are really no different and are the subject of the next chapter.

CHAPTER NINE

THE DOMAINS OF ROYALTY: CASTLES AND PALACES

A consideration of the visual persuasiveness of castles and palaces is a natural next step from assessing governmental buildings. The architecture of castles and palaces differs little in purpose from that of governmental buildings. Power, authority, respect, control, iconicity are also the visually persuasive messages of the domains of royalty. If there is something that may make castles and palaces a bit different from government buildings in general it is their association with royalty. For the absolute ruler of a country, and in some cases the appointee of the Deity, there must be every evidence of the greatest wealth in the land. It goes without saying that monarchs intended their castles and palaces to be splendid and luxurious. Nowhere is this principle more evident than at Versailles, southwest of Paris, and the palace of Louis XIV the Sun King. The extravagance of Versailles set the standard for almost every palace built after it. Coincident with splendor, there is also the need to evince power and dominion. Size, solidity, and location are the usual means of accomplishing this goal, and both castle and palace grounds are ordinarily expansive or elevated.

What is the difference between castles and palaces? Ordinarily, a castle was a walled fortress, often surrounded by a moat, and containing several residential and administrative buildings, storage facilities, a dungeon, and a palace—all the necessities required for sustenance should there be a prolonged siege. Since castles were fortresses, their sites were usually on a high promontory and their fortifications extensive and massive. Unlike palaces, castles were not necessarily visually ornate or splendid. Palaces, by contrast, were the residences of monarchs and perhaps the centerpieces of castles. Sometimes, however, palaces were not enclosed by walls or part of a castle complex. Palaces usually differed from both castles and government buildings by lavish appointments within and the housing of a variety of art objects and a cadre of servants. In this chapter, castle refers to a walled complex, while palace refers just to the monarch's residence. A convenient beginning point for this examination of castles is Great Britain and Durham Castle, because it has already been described in Chapter Six as the present home of Durham University.

Castles of Great Britain and Continental Europe

To note again those architectural features which convey Durham's distinctiveness, one would begin with the castle's location on one of Durham's highest hills and in a bend of the River Wear. Though no fortress is truly impregnable, it would have been very hard indeed to cross the river, scale the heights, and penetrate or climb the walls of Durham Castle. Its visual impact is that of a solid, sound, elevated structure, whose inhabitants were safe and secure. That safety and security remains as a comfort to those students today who are fortunate to be its residents. Durham Cathedral, standing across a green, appears to bless the castle's inhabitants by its proximity. Although it is not a castle, the walls still standing in the city of York, south of Durham, are a reminder of the fortified or walled cities of earlier times and the protection they provided. Visually, such walls argued that this was not a place to consider as a target, and these were people not to be aggressed upon.

Fig. 9-1 White Tower

Although it may not always have been so, the castle complex referred to collectively as the Tower of London is today's quintessential castle. Sitting on the banks of the Thames River, Her Majesty's Royal Palace and Fortress is a huge walled complex encircled by a broad moat, although the moat is no longer functional as such. The walls are tall, thick, and built of forbidding looking gray stone. The complex dates to the 11th century and the time of William the Conqueror, but it has been modified, expanded, and altered for different purposes over the centuries. It is perhaps most commonly thought of as a prison, since so many monarchs incarcerated their enemies and potential enemies there. “Sent to the Tower” usually was a phrase that portended the ultimate banishment if not outright death.

Today, there remains the executioner’s block on which many well-known English men and women, including wives of kings, bent their heads to the ax.

Within the walls, the most distinctive building and the one which gives the complex its name is the White Tower. The architecture is Romanesque or Norman, and the structure is a simple square with turrets at each corner and square battlements along the tops of the walls. The whitish gray stone of the structure gives it its name, and it was at an early date whitewashed (Field 2007). The tower is a keep or dungeon, although it is large enough to include quarters for the monarch, a small chapel, and an undercroft for storage. The visual impression of the straight lines and square as well as the tower’s vertical dominance within the walls is one of mass, strength, and power. Among the other buildings within the walls is the Waterloo Barracks, wherein are displayed the Crown Jewels.

Fig. 9-2 Edinburgh Castle

Edinburgh Castle, like Durham Castle, is located on the highest point of the city overlooking the Firth of Forth and the city itself. It stands at the upper end of the Royal Mile. Like both Durham Castle and the Tower of London, this is a rugged and forbidding stone complex. It has been described as appearing to have grown out of the volcanic rock on which it sits (Field 2007). The architecture is neither ornate nor particularly visually interesting, so it is the mass, the location, and the general impression of strength that makes the structure compelling. The buildings are primarily Georgian in style, constructed in square-like shapes of heavy stone. The texture of the walls is rough.

There are the usual barracks, chapel, and storage facilities as well as cannon displayed prominently. The Royal Palace is within an interior square of the complex and features a single tower and battlements above the entryway. It was the birthplace of James I of England (James VI of Scotland), the son of Mary, Queen of Scots, and today is home to the Honors of Scotland and the Stone of Destiny or of Scone. Many of the castle's buildings today serve as museums of one sort or another, including the National War Museum, the Regimental Museums, and the Scottish National War Memorial (Tabraham 2008).

As mentioned above, one tends to think immediately of Versailles when French castles and palaces are mentioned. However, there are at least three other complexes which perhaps better represent the notion of the French castle. The first is the predecessor of Versailles, the Château de Vincennes, on the eastern edge of Paris, while the other two, Carcassone and Aigues-Mortes, lie far to the south. Although not nearly so popular with visitors as Versailles, these three have become successful enough as tourist attractions to assure their preservation. All three are remarkably well preserved, perhaps because of their sturdy stone construction.

The Château de Vincennes was built by Charles V during the period 1361 to 1380 CE (http://en.chateau-vincennes.fr). The layout of the castle, like that of the Louvre in Paris which it predates, is a simple rectangle. The walls and the area within, called the *enceinte*, are surrounded by a moat that was 27 meters wide. Nine towers were located along the sides and at each corner. Within the castle walls, the most visually prominent structure is the donjon, which is a square tower whose sides are 50 feet in length and which rises six floors in the air. Round turrets are located at each corner. The King's study was on the second floor.

Directly across from the donjon is the Sainte-Chapelle, which resembles the one on the Isle de la Cité in Paris discussed in Chapter Five. It is of Gothic design with a single entry under a large rose window. Architecturally, it is probably the mass and powerful vertical lines of the donjon that give the Château de Vincennes its principal visual force. However, the unadorned horizontal lines of the walls and moat provide an impressive strength, and the chapel is certainly distinctive in its Gothic adornment and colorful stained glass. Far to the south of France, Aigues-Mortes and Carcassonne match the visual power of Vincennes. Both are stone walled fortresses in much the same pattern.

Fig. 9-3 Vincennes

Aigues-Mortes lies close to the Mediterranean Sea near the marshlike Camargues area. Built in the 13th century as essentially a port town, it is a simple rectangle nearly the size of Vincennes. However, the walls of Aigues-Mortes enclose an entire city rather than simply a number of buildings, a chapel, and a donjon. The most visually impressive aspect of the walled city is its powerful solidity. The crenellated, notched walls are stark, unadorned stone with five towers and ten fortified portals. Again, there are strong horizontal and vertical lines. As a fortress should, Aigues-Mortes conveys rugged mass (http://www.provencebeyond.com/villages). Similarly also does 11th century Carcassonne in the Languedoc region northwest of Aigues-Mortes.

Fig. 9-4 Aigues-Mortes

Carcassonne was rebuilt in the 19th century, thus it looks like a movie set (http://www.frenchentree.com/languedoc-roussillon-holiday). Indeed, *Robin Hood, Prince of Thieves* was partially filmed there. Carcassonne is also a walled city. It stands on a steep hilltop above the Aude River and the newer city. There are inner and outer crenellated walls. The outer walls have 14 towers, while the inner ones feature 24. Most of the towers feature conical turrets, and the Narbonnaise Gate is flanked by two closely spaced towers with conical turrets. Within the walls, there is a castle and the Gothic Basilica of Saint Nazaire. The hilltop location, the crenellated walls, and the towers combine to make Carcassonne a compelling sight indeed.

There are, of course, castles and fortresses throughout Western Europe, but space considerations limit the remainder of this discussion to three very visually distinctive others: Hohensalzburg, the Austrian fortress and castle; Prague Castle; and the much smaller Castel sant'Angelo, or Mausoleum of Hadrian, in Rome. The first, Hohensalzburg, greets the visitor arriving from any direction in Salzburg. Salzburg is itself surrounded by snow-topped mountains, but within its valley setting a tall bluff, called the Festungsberg, rising above the Salzach River is the site for the fortress and castle.

Fig. 9-5 Carcassonne

Fig. 9-6 Hohensalzburg

Hohensalzburg is either the largest fortress and castle in Continental Europe or one of the largest depending on which source one consults. It was begun in the 11th century, but it took a number of centuries for it to reach its present size and appearance. The crenellated fortress walls feature a small number of towers but overall give the impression of simple straight lines. The stonework is off-white, although the walls glow orange as the sun sets. Within the walls, there is a church and a courtyard ringed with a number of small, shop-like buildings. Architecturally, Hohensalzburg is a visually compelling site, first and foremost, and has impressive size and clean lines. For many of the same reasons, Prague Castle is also architecturally compelling.

Although not set in the midst of a mountain range, Prague in the Czech Republic is a hilly city, and the highest point is the site of Prague Castle and St. Vitus Cathedral discussed in Chapter Four. Like Hohensalzburg in several ways, Prague Castle is a frequent subject of postcards from the city, it is easily visible from anywhere in Prague, and it is massive. It is not a fortress, although there certainly are walls within the complex, but rather a series of interconnected governmental and ecclesiastical buildings which surround a sharply narrow triangle of interior space. More buildings lie parallel to the main complex across the principal access road. Dating to the 10th century, Prague Castle also claims to be the largest in Europe.

Fig. 9-7 Prague Castle

The final castle to be considered in the present discussion is not technically such a structure in spite of its name. Castel sant'Angelo, on the banks of the Tiber River in Rome was originally built in the 2nd century to be the burial site of the Emperor Hadrian. Later the building was used by the church and as a museum. Although not on a hill, the building's location on the right bank of the Tiber is reached by crossing an imposing bridge, which draws visual attention to the castle. The castle itself is a perfect cylinder with tower wall on one side. Although smaller than the other castles examined here, Castel sant'Angelo certainly fits the general pattern of a defensible, protected, and massive site with imposing walls.

Fig. 9-8 Castel sant'Angelo

Although they were sometimes located within castle walls, palaces are primarily interesting as types of visual persuasion, because they were royal residences. In the following chapter, the visual significance of residences in general, including apartments and hotels as well as homes, will be considered. One, perhaps primary, function of a residence other than to provide shelter was to lend status to the owner or owners. In the case of palaces, lending status to the owner was paramount, and it was almost wholly visual. Like castles, palaces were huge and located propitiously. Great size was important to provide for the needs of the royal court and for visitors, but it also seems to have been an end in itself. As in the case of many other types of architecture, size or mass equaled power and significance. Unlike castles, palaces were magnificently decorated both on the buildings themselves and in the surroundings, which were often the equivalent of game preserves or parks. Ornamentation, color, and geometric designs permitted not only a compelling vision but also gave royalty the opportunity to display the splendor which only the most lavish coffers could provide. In a word, palaces were supposed to be stunning. There is probably no more stunning a palace in the world than that of Louis XIV at Versailles southwest of Paris, so that is the proper place to begin this examination.

Palaces of Continental Europe and Great Britain

Unlike the case of castles, which are relatively similar and can be characterized architecturally with only a few elements, palaces are complex blends of many visual features. What follows is, therefore, a cursory account at best. The Château de Versailles is the most prominent structure in a complex, which includes the smaller Grand Trianon and Petit Trianon palaces and vast, ornate gardens with incorporated pools, fountains, and statuary. The vast gardens are the consequence of Versailles originally having been a hunting lodge built by Louis XIII in 1624 (Magi 1990). Beginning in 1661, Louis XIV began a remodeling and expansion of the hunting lodge, and over the next two decades he and the French court moved to Versailles.

From 1682 until the French Revolution in 1789, Versailles was the center of French government and home to both Louis XV and Louis XVI. The approach to the main palace is through a massive iron gate painted in black and topped with gold finials. The gate is flanked on either side by complementary black fencing common to most of the prominent châteaux now belonging to the French government. Beyond the gate is a vast stone paved courtyard, which slopes gently upward to the palace and to a huge statue of the Sun King astride his horse. The main palace structure is in the shape of an inverted U, much like the Louvre. Flanking the approach along the courtyard are wings which continue the lines of the inverted U and which in turn are flanked by rectangular buildings set at a 90 degree angle to the main one.

Fig. 9-9 Versailles

The architectural style of the palace is French Baroque. The main building has three stories and a gently curving roof. Paired Doric columns decorate the first floor entry. Arched windows dominate the second floor interspersed with sculpted busts, and the third story continues the theme of the second without the sculpture but with a balustrade. Finally, the entry is topped with a clock flanked by ornate sculpture. The ornate theme of the entry is continued within by broad staircases and, preeminently, the Hall of Mirrors. In this vast rectangular hall, tall mirrors in gold frames line both sides of the room and Baroque chandeliers hang from the ceilings. The main palace was so ornate and vast that Louis XIV had a smaller palace built across the gardens to the northwest called the Grand Trianon, where he spent time with his mistress. Across the gardens, the smaller Petit Trianon was a favorite of Marie Antoinette.

At nearby Fontainebleau to the east is located another splendid royal château. Its architecture is not so ornate as that of Versailles, having predated the emergence of the Baroque, but the estate is comparable in size. It is also surrounded by a vast hunting park, which bespeaks its similar origin as a hunting lodge built in the 12th century by Louis VII (Samoyault 1994). François I in the 16th century is responsible for converting the hunting lodge to a true palace, and subsequent monarchs remodeled

Fig. 9-10 Hall of Mirrors

and expanded the complex. François was strongly influenced by the architectural style of the Italian Renaissance, and Fontainebleau was remodeled accordingly (Glancey 2006). The main entry is at the end of a wide cobblestone street. The building is three stories high and has repeated towers along the façade. Its most impressive feature is a curving horseshoe staircase leading to the second floor. Stately rather than ornate, the building is simple stonework with windows on each floor separated by Doric pilasters.

Fig. 9-11 Fontainebleau

Both Versailles and Fontainebleau are visually compelling because of their sites, their size, and their architectural styles. Palaces, unlike other buildings however, are also influential through the opulence of their interiors. Room after room, especially those assigned to the royalty themselves, open on richly brocaded walls, elaborate chandeliers, ornate flooring, lush draperies, and expensive paintings and sculpture. The Hall of Mirrors in Versailles is perhaps the single most opulent and impressive room in any palace, and it is no wonder that it was the scene of the signing of the Treaty of Versailles ending World War I. Many rooms at Fontainebleau are typical of those in palaces all over Europe. The Trinity Chapel, for example, is primarily white with gold accents and has a domed ceiling covered with paintings. The Queen's bed chamber is heavily ornamented with gold railings and wall decorations over a silk brocade wall. The ceiling is carved blue and gold (Samoyault 1994).

Neither so large nor so richly ornate, the Royal Palace of Belgian royalty in Brussels is nonetheless a visually impressive sight. Located in the Upper Town of Brussels overlooking the older city, this palace stands at the top of the highest point of the city diagonally across from the country's finest museums of art. Facing the palace is a cobblestone square called Royal Place. Centered behind this square is the large Neoclassical structure of the Royal Palace. Although there were previous royal structures situated here as early as the 10th century, the palace of today dates only to a remodeling in the late 19th century. Like Versailles and Fontainebleau, the Belgian Royal Palace is surrounded on two sides by a park, the Parc de Bruxelles. The building is primarily a long horizontal expanse, which curves to the rear at each end. Its entryway is a Classical

Greek portico of eight Corinthian columns topped by a pediment and sculpted tympanum. Two wings project outward at each end of the façade and are also decorated with Corinthian columns. East of France and Belgium lie Germany and Austria, where there are a number of remarkable palaces. Surely the most visually compelling of those is Schloss Schönbrunn in Vienna. Quaint, but in their own way also visually compelling, are Hohenschwangau and Neuschwanstein, both of which are located in Germany south of Munich near the Austrian border town of Füssen.

Fig. 9-12 Belgian Royal Palace

Located in a suburb of Vienna, Schloss Schönbrunn is a palace in the manner of Versailles and Fontainebleau. The building is quite large, and a large park-like formal garden with pools and statuary extends to the rear and leads to a hillside military monument facing the palace called the Gloriette. A small complex with a residence and grounds called Katterburg came into the possession of Maximilian II, of the Hapsburg dynasty, in 1569 (http://www.guide-to-castles-of-europe.com). The emperors Matthias and Ferdinand II used the palace and grounds for hunting, and it eventually passed to the Empress Maria Theresa, who had it rebuilt in the mid-18th century as the magnificent structure it is today. The Emperor Franz Joseph was born in the palace in 1830 and used it as his residence when he rose to power. Mozart visited and performed at Schönbrunn as a child (Ellmerer 1993). The palace once housed some 1500 people.

Fig. 9-13 Schönbrunn

Although not set on a hill, Schönbrunn is architecturally imposing by its location at the head of vast grounds. The building itself is also quite large and is a sort of restrained Baroque piece of architecture. It is a principally horizontal building of three stories, although wings project out at each side and another wing projects from those. It is of yellow stone and has a façade decorated with Ionic columns and pilasters and topped with a balustrade running along the base of the roof. Curving staircases on either side of the first floor entry form a horseshoe leading to the second floor. The interior is highly ornate in the manner already established here. Franz Joseph I's study is comparatively restrained. It is done in plum wall covering with complementary draperies and has portraits hanging on some of the walls. Other rooms, such as the Round Chinese Room, are as gaudy as any rooms in Europe. This room, for example, is done in white with gold accents and a massive crystal chandelier. The Great Gallery resembles Versailles's Hall of Mirrors and features a frescoed ceiling, wall sconce lighting, and massive chandeliers.

Fig. 9-14 Neuschwanstein

Neuschwanstein is the palace of Ludwig II of Bavaria built in the late 19th century and the model for Sleeping Beauty's Castle in the Disney theme parks. The architecture is basically Romanesque Revival, but there are also elements of Gothic. The palace is principally a multi-story basilica with added balconies and extensions along the rear. It has a comparatively plain ochre stone gatehouse, while the main building is of off-white stone. Turreted towers of different heights are located along the roofline and are the most memorable aspects of the design.

Neuschwanstein is situated on a peak above the valley in which lie Schwangau and Füssen, and it occupies virtually all of the hilltop. For this reason, the palace is usually photographed from a distance to allow for the full visual effect. From a distance, the structure does in fact live up to its popular designation as "the fairy tale castle." Within, Neuschwanstein is the equal of any palace in Europe. The Throne Room, for example, is richly ornate. Along the sides are Corinthian columned arches rising to a balcony which repeats the arch theme. Every surface is gaudily painted in reds, golds, and greens. Some of the painting is of geometric shapes, and the rest depicts religious themes.

In the valley directly below Neuschwanstein and within walking distance is located Hohenschwangau, the early 19th century palace in which Ludwig II grew up. This structure is much less ornate and resembles a fortress as much as it does a palace. A Gothic structure of yellow stone with sienna tile accents, Hohenschwangau consists of two primary connected buildings. The tallest of these is a multi-story square tower with smaller towers at each corner. A smaller, shorter rectangular building features two turrets along one side. The towers have crenellated parapets. The interior, in such rooms as the Hall of the Swan Knight (Lohengrin) and the Queen's Living Room are as richly decorated as any of those in Neuschwanstein.

Fig. 9-15 Hohenschwangau

Further to the east lies the Italian city of Venice and its iconic Doge's Palace. Venice was for centuries a state by itself, the Republic of Venice, whose ruler was an official elected for life called the Doge or Duke. Located directly on the Venetian lagoon in front of the Basilica of San Marco, the palace was a fortification like a castle, the residence of the Doge, and a hall of justice. Prisoners adjudicated in the palace could be taken directly to an adjacent prison over the Bridge of Sighs which will be examined later//www.italyguides.it). Architecturally, there is surely no more prominent site for a building in Venice than adjacent to the piazzetta and fronting the lagoon.

The architecture of the Doge's Palace has been called Venetian Gothic and dates to the 9th century (http://www.italyguides.it). However, there is no Gothic building outside Venice that is quite comparable to the Palace (Glancey 2006), and its architecture clearly prefigures the Italian Renaissance style. There are three stories, the lower two of which have loggias bound by graceful, narrow columns and arches that keep the comparatively small and squat building from appearing stolid. The third story is plain stone like the walls of a castle with short, arched windows. The roofline is ornamented with triangular peaks like a tiara. The predominant color of the palace is white, but there are light pink accents. Within the palace, there is a courtyard and stairways to the upper floors. The courtyard repeats the loggia theme of the exterior, but there are many alcoves on both of the two lower stories with sculpture. Many of the rooms within the palace are decorated with dark wood, gilding, and painted ceilings.

Fig. 9-16 Doge's Palace

In Spain, the Palacio Real on the western edge of Madrid is a splendid mid-18th century example of royal extravagance. It is the largest palace in Europe (http://archiseek.com) at 135.000 square meters of space. The location is a low hill, and it is part of the complex which includes the Cathedral of Almudena. The Palacio Real faces the Cathedral across the vast stone Plaza de Armas, which is flanked on each side by loggias and offices. The building's architecture is Baroque, although Neoclassical elements are clearly evident. Primarily of white stone, the multi-story palace has a façade with Corinthian columns and pilasters, a balustrade along the roofline, and a central dome. The Palacio Real is as ornate within as any palace examined here, but it is distinctive in that it contains several pieces of museum-worthy art. The Guard Room and Throne Room both have ceiling frescoes by Tiepolo. The Hall of Columns has tapestries woven after designs by Raphael, and there is a Stradivarius Room with 300 or so violins by the Italian master.

Fig. 9-17 Palacio Real Madrid

The Hungarian Royal Palace in Budapest, also known as Buda Castle, sits at the top of a crest overlooking the Danube and the area of the now combined city known as Pest. This architecturally compelling site is also diagonally across the river from the Hungarian Houses of Parliament reached via the Chain Bridge. The palace location dates to the middle of the 13th century, but the present complex is the product of the mid- to late-19th century. The architecture is essentially Baroque, although many modifications over the centuries and a rebuilding after near total destruction in World War II mean that there are a variety of architectural influences. The main structure is five stories high, the ground floor of which is rough dark stone. The upper walls are of smooth stone. Along the façade, the ground floor has arched windows and an entry portico. To each side of the entry, the building projects outward. The second and third floors feature balconies ornamented with Corinthian columns and engaged columns. The fourth story is topped by a balustrade, while the upper story is decorated with a balustrade. A central dome rises behind the entrance. There are no longer many interior spaces to indicate the splendor of the earlier years, since today the palace is occupied by the National Gallery, the National Library, and the Budapest History Museum. However, the ballroom appears to have been designed with Versailles's Hall of Mirrors in mind.

Fig. 9-18 Buda Castle

Hungary is no longer a kingdom, which explains the conversion of Buda Castle into a group of museums. Sweden, however, remains a parliamentary monarchy much like Belgium, Spain, and the United Kingdom. The Stockholm Royal Palace in Gamla Stan on the island of Stadsholmen is still the residence of the Swedish king and his family. The palace forms a square with an interior courtyard and wings at each corner. It is built of brick with a sandstone facing. The architectural style is Baroque but with strong Neoclassical elements. A balustrade encircles the roof of the three story palace. The exterior walls are gray and white. Pilasters and arched windows adorn the first and third floors, while sculpted pilasters line the second floor. On an elevated portion of Gamla Stan, the Stockholm Palace dominates any approach to the Old Town. It site, size, and restrained Baroque architecture are the architectural elements which give the palace its visual power.

Fig. 9-19 Stockholm Royal Palace

Finally in this account of the domains of royalty, there are two palaces in the United Kingdom both of which belong to the British monarch. The Palace of Holyroodhouse in Edinburgh, Scotland, is now primarily a museum, although it is a working residence to which the Queen and her family come for regular visits. In London, Buckingham Palace, the official residence of the Queen, is one of the most familiar sights in the Western world. The Palace of Holyroodhouse is situated at the foot of the Royal Mile in Edinburgh rather than on a hill, but juxtaposed with Edinburgh Castle at the other end of the Royal Mile the site is impressive. The grounds set the palace off from the surrounding buildings, and of course today the Scottish Parliament sits across the street as it were.

Buildings on the Holyrood site date to the 13th century, when an abbey was built there. The present building was erected in the 15th and 16th centuries (http://e-architect.co.uk). The architecture is Baroque, although aspects of the building show Norman influences. The corners of the building, for example, are round towers topped with conical turrets and crenellated battlements. The stonework is rough and lacking in color. The main entry into an interior courtyard is a single portal framed by paired Doric columns rising two stories. The royal coat of arms is sculpted on the wall above the door, and there is an octagonal clock tower which arises above. The roofline is encircled with a balustrade. The interior courtyard is rather more ornate than the entryway. The ground floor is an arched loggia featuring Doric engaged columns. The second floor alternates windows with bare walls, while the third floor features windows separated by

Corinthian engaged columns. Although one might expect the interior of Holyroodhouse Palace to be rather more austere than rooms in other palaces, given the stereotype of the frugal Scot, the decoration is quite elaborate. Mary Queen of Scots' bedchamber, for example, is highly ornate with rich blues, greens, and golds.

Fig. 9-20 Holyroodhouse Palace

From the days of the British Empire, through the Nazi Blitzkrieg of World War II, to the worldwide attention given today to the British monarchy, Buckingham Palace has been the focus of both news and photography of the city of London and one of its enduring icons. A viewing of the changing of the guard in front of the palace is high on the list of most visitors' things to do. To be sure, Buckingham Palace is easily in the same category as the most important palaces that have been considered here, such as Versailles and Schönbrunn. Located centrally in London within walking distance of Victoria Station, the palace is surrounded by parks: Green Park, St. James's Park, and Hyde Park as well as its own gardens. Adding to the visual prominence of the site is a large square outside the palace forecourt containing a large monument and statue memorializing Queen Victoria. A wide mall leads away from the square to the mobilization point for the Royal Guards. It is in the forecourt that the changing of the guard takes place. This area is surrounded by fencing and gates of black iron topped with gold finials.

Originally, Buckingham Palace was simply the residence of the Duke of Buckingham, but it was acquired by King George III in the 18th century and remodeled by John Nash in the 19th century. As such, its architectural style is Neoclassical, Georgian, or Palladian. It was first occupied as the official residence of the British monarch by Queen Victoria. The façade of the East Wing of the palace is the most familiar view, since it is before this that the changing of the guard takes place. The East Wing opens onto an interior courtyard which leads to the part of the palace containing the principal residential and ceremonial rooms. The east façade spreads out on either side of a single arched stone entryway topped with Corinthian columns and engaged columns and a Classical Greek pediment. At each end of the façade, the building projects forward a bit and reveals a matching façade.

Fig. 9-21 Buckingham Palace

Although much of the interior of the palace is off limits to visitors, the rooms that are available to view are spectacular in the vein of Versailles and Schönbrunn. The Grand Staircase to the State Rooms is horseshoe shaped and done in white and gold. The Blue Drawing Room is ringed with marble columns, hung with elaborate chandeliers, and decorated on its ceilings with multiple geometric figures. The State Dining Room has a white ceiling with many gold accents. The walls are maroon along with the carpet, and gilt framed paintings of previous royals adorn them. The Throne Room is decorated similarly.

This chapter has been an extension of Chapter Eight, concerning government buildings, with the purpose of examining the castles and palaces of royalty. Fundamentally, these places are also government buildings, although they are also usually residences. Like government buildings, the domains of royalty exude power, authority, dominion, and control. Unlike government buildings, although the difference may only be one of degree, castles and palaces are opulent thus using great wealth to signify power and authority. In the next chapter, ordinary and perhaps some extraordinary places of residence will be studied. As it turns out, the absence of royalty does not alter the fundamental means by which people seek to influence others. One does not merely seek shelter when buying or building a home, or an apartment, or when selecting a hotel while away from home. One also seeks the effect and influence of visually persuasive architecture.

CHAPTER TEN

PLACES OF RESIDENCE: HOMES, APARTMENTS, AND HOTELS

The idea of a home occupies a central niche in the collective psyche of the Western world, at least in those countries where private property makes it possible to express oneself in a place of residence. The idea of a home transcends the mere notion of a house. A house is only a structure, while a home is imbued with the allied ideas of family, relationships, security, warmth, and a host of other occasionally intangible ideas. "Home is where the heart is" goes the popular expression. Home, therefore, is a psychological construct about which home owners have a variety of understandings. Surely one of the most prominent of those understandings is that one's home is iconic for its owners much as museums are culturally iconic. Homes are status symbols, and many owners feel differently only in degree from the monarch who sought to demonstrate his power and authority through the expense of great wealth. The locations of homes are as well ways of demonstrating status, hence the presence in virtually every city of enclaves of the rich and the knowledge of most residents of desirable addresses.

Homes, as they are thought of conventionally, are of course not the only common places of residence in any part of the world. There are, for example, apartments, which range in size and complexity from simple bungalows in the poorer parts of most cities to large, luxurious spaces in the high rise buildings of cities like Chicago and New York City. Although they are normally only temporary residences, hotels are also significant structures in considering people's living places. This tripartite division of places of residence is a simplification of the structures one might consider as examples of compelling architecture.

Additionally, the range of structures within each type is so vast that it is possible to consider only a few examples within the range. Whole books, for instance, have been written on American houses alone. Blumenson's (1977) guide to American architectural styles and the McAlesters's (1985) field guide to American houses are two especially valuable examples, which the student of places of residence will want to consult. Although these two books do not address the issue of architecture as visual persuasion, both are richly illustrated. This chapter will examine an eclectic mix of homes starting with some residences which are nearly palaces. It will conclude with a brief reference to apartments and to the role hotels play as temporary places of residence for travelers.

Homes

In the US, some of the most visually compelling places of residence are the well-known homes of American presidents. These include George Washington's home at Mount Vernon, Virginia, Thomas Jefferson's home at Monticello, Virginia, and the temporary home of every president, the White House in the District of Columbia. Akin to these presidential homes is the Iolani Palace, which was the home of the kings and queens of Hawai'i. The presidential homes and the Iolani Palace were the centers of large estates and were designed not only for their visual power but also to

Fig. 10-1 Mount Vernon

accommodate large numbers of guests and visitors. The White House, of course, is not only a residence but a complex of government offices and spaces that functions much as museum galleries do.

George Washington was a wealthy planter in addition to his service to his country as a military general and the nation's first president. As a result, Washington's Mount Vernon is a 500 acre plot with room for fields of wheat, corn, and other grains, which grew over the years of Washington's residence into an 8,000 acre estate. It was originally planted with tobacco, but tobacco farming was abandoned when it was found that the fertility of the land had been depleted from previous tobacco planting (Rees 2001).

Washington's home itself sat on a bluff overlooking the Potomac River, a location that not only provided ready access to travel by boat but also was visually compelling. It commanded a view in all directions, and it was an impressive sight to anyone traveling along the river. Today, tours of the estate may be arranged which bring the visitor to Mount Vernon from the river.

The architecture of Mount Vernon, designed by Washington himself, was influenced by Andrea Palladio, whom Washington admired. The final building plan might best be described as Georgian in style. The building is three stories of pine wood construction painted white and rusticated to resemble stone. The roof is red shingle, so that the color of the structure is quite eye catching. Two chimneys, two dormer windows, and a central cupola adorn the roof. The entry façade is plain but perfectly symmetrical following Palladio's principles and features regularly spaced windows and shutters on both of the first two floors. The rear of the home, which faces the Potomac, is more ornate in that it boasts a portico with eight square columns and Doric-like capitals.

From each side of the house, curved colonnades lead to two outbuildings. These columns resemble those on the rear portico and are connected with delicate archways reminiscent of those common in the Italian Renaissance. Before the front of the home, Washington installed an elliptical courtyard surrounded by a circular driveway. Posts connected by chains surround the ellipse and provided a place for horses to be tied up. In the middle of the courtyard is a sundial. From the standpoint of visual persuasion, the site of Mount Vernon is primary. Its Palladian symmetry, its use of rhythmically repeated elements such as windows and columns, its color, and its size combine to create a most visually compelling structure.

The interiors at Mount Vernon are, in most respects, reflections of the decorative restraint and simple symmetry of the exterior. Two rooms, however, depart from this restraint with highly ornate and colorful features. The large dining room is wallpapered in verdigris green with white trim along the cornice boards. Above the cornices, the wall curves upward to the white ceiling and is decorated with a white draping chain motif. The ceiling is bas-relief. On one side of the room is a fireplace with two Ionic marble columns on either side, while across from the fireplace is a highly ornate Palladian window. The central portion is an arched window, and this is flanked with rectangular windows set off on each side by white pilasters whose centers are green with white applied designs. The second rather ornate room is the smaller West parlor, which is done in blue.

Fig. 10-2 Monticello

Thomas Jefferson, author of the Declaration of Independence, founder of the University of Virginia, and the nation's third president, was also his own architect and was strongly influenced by Palladio. He owned five editions of Palladio's work on architecture. Like Mount Vernon, Monticello is the focal point of a 5,000 acre farming complex. Like Washington, Jefferson thought of Monticello as his retreat from the politics which he hated (Stein 1997). The house itself sits at the top of a hill, the top of which Jefferson leveled to serve as the foundation. The structure began as a single room residence which Jefferson occupied as a bachelor and as a newlywed. The present building is the result of gradual expansion over many years, a time interrupted by several absences from Virginia during periods when Jefferson served in a variety of governmental capacities.

In spite of having been influenced by Palladio, Washington's Mount Vernon is not so clearly in the Palladian Neoclassical style as is Monticello. The home is of deep red brick with white columns and trim. There is also a dome above the entrance portico, a balustrade which surmounts the roofline, and a smaller balustrade along the peak of the roof. The façade is absolutely symmetrical with white trimmed windows flanked by deep green shutters ranging from side to side. The entrance is a portico with gently sloping stairs and a Greek entablature and pediment sitting atop six Doric columns. The dome has an octagonal base around which are equally spaced round windows. Before the entryway is a large lawn suggestive of the one at Mount Vernon. The interior of the house is rather less ornate than some of the rooms at Mount Vernon, with most of these spaces done in pastels with straight lines. Site, symmetry, color, and rhythm are the principal components of the visual message of Monticello.

It is difficult to imagine a more familiar sight than that of the White House in Washington, DC, or a more familiar address: 1600 Pennsylvania Avenue. This building, which is an office complex as well as a residence, accommodates each President of the United States and his or her family during a four or eight year term and contains numerous rooms for entertaining visiting dignitaries. The cornerstone for the mansion was laid in 1792, and the construction was overseen by George Washington. Not finished until 1800, however, Washington did not ever live in the White House. John Adams, the second president, was the first resident. Like Mount Vernon and Monticello, the White House was altered and enlarged over the years by successive presidents and survived a fire which nearly destroyed it during the War of 1812 (http://www.whitehouse.gov). The building is surrounded by large lawns and fountains on both the south and north sides. The surrounding greenery is a colorful complement to the stark white sandstone walls of the building.

The architectural style in vogue at the time of the building of the White House was Neoclassical, and the structure is a beautiful example of that style. Both the north and the south façades are visual icons for Americans and many others around the world. It is the north façade that is the normal entryway for visitors, and it is two stories high. There are eight ranks of windows on either side of an entry portico. The first floor windows each have curved pediments, while the second story ones are flat. Within the colonnaded portico, three more windows on each floor repeat the pattern of the ones on each side, giving the symmetrical north façade a rhythmic visual effect. The columns are Ionic and are surmounted by a Classical entablature and pediment. A small balustrade runs along the roofline.

Fig. 10-3 The White House

The south façade is similar to the north one, but it has three stories showing. Additionally, there is a circular portico flanked by semicircular stairs on the first floor and featuring six Ionic columns on the second and third floors. Some of the interior rooms of the White House are as famous as the residence itself. In the West Wing, there are offices for the presidential staff and the president's own Oval Office. Within the main building, there is the Blue Room, where the president ordinarily receives official visitors, the multipurpose East Room, the Green Room, a sort of state parlor, and the Red Room, used for a variety of purposes over the years (http://www.whitehouse.gov). The White House's visual impact is primarily the result of such familiar architectural features as its site, its symmetry, its color, and its rhythmic pattern of columns and windows.

Another dramatic place of residence quite unlike the architecture of the houses above is the Iolani Palace in Honolulu, Hawai'i, the fiftieth state of the US. While one might expect to find an exotic structure in the tradition of Pacific Islander architecture, the Iolani Palace instead is an Italian Renaissance building. The four story palace was built by King David Kalakaua between 1878 and 1882. The royalty of Hawai'i, however, lived in the palace only 11 years, since Queen Liliuokalani was removed from the throne in 1893 by US Marines acting at the behest of US commercial interests in the island nation. The palace is today a museum.

Fig. 10-4 Iolani Palace

The Iolani Palace is located in the central area of the city of Honolulu and is surrounded by outbuildings, a large lawn, and a number of banyan trees. Its site is quite dramatic and is also surrounded by iron fencing. The façade of the building is decorated with stairs leading from ground level to the second story. The first story is half below ground level. There is a central portico extending all the way to the roof ending in a tower with four arched windows, and this portico features two paired columns on either side of central archways. The second and third floors are ornamented with four slender Corinthian columns on either side of the central portico which frame horizontal porticos. At either end of the building, there are solid wings of limestone.

As might be expected, the Iolani Palace interior is no less dramatic than the exterior. The main floor—the upper floor was residential—is done with highly polished dark native Hawaiian wood and off-white walls. Arched doorways lead off the Grand Hall to a Throne Room, a Blue Room used for receptions, and a State Dining Room. There are elaborate chandeliers and a dramatic dark wood stairway leading from the Grand Hall to the second floor.

Although neither presidents nor royalty, many great figures of American industry demonstrated their wealth by building visually imposing homes in especially prestigious locations in major American cities. Indeed, they often used the building of a home as a type of competition. A single but representative example of this type of home is that of Henry Clay Frick in Manhattan. The home is located on Fifth Avenue at 70th Street. Frick is said to have built his home in 1913-1914 to outclass the home of his rival, Andrew Carnegie, located about a mile north. Frick became wealthy through coke mining and sales and had financial interests in railroads and steel production. He was also a devotee and patron of the arts, and his private collection of paintings and sculpture is housed today in his home much as it was during his lifetime. The home is now a museum.

Fig. 10-5 Frick Mansion

The site of the Frick Mansion is clearly prime real estate in one of the most expensive locations in the US if not the world. Not only is the site architecturally significant, the structure occupies most of a city block. It is built of gray limestone and conveys the visual impression of vast size. The side of the building facing Fifth Avenue sits behind stone and wrought iron fencing. It is three stories tall. The uppermost half-story has rhythmically repeated rectangular windows and is lined with a balustrade. The first and second stories also feature windows, both rectangular and arched, and in the center of the building are four two-story Ionic pilasters. Within the building is a courtyard and pool. Today as a museum, the interior is especially impressive, although not well lit. Living in the mansion surrounded by such art as can be found in the collection must have given Frick great pleasure and should have reassured him of his financial status.

Although not captains of industry, the leaders of the early settlers of America in such places as Alexandria, Virginia, sought also to establish their prominence in the community through the construction of elegant homes. Among the architecturally impressive residences in what is now referred to as Old Town Alexandria is the Carlyle House, a beautiful example of Palladian symmetry in the vein of Scottish and English manor houses. Carlyle House was built by a Scottish merchant for his new spouse in 1753 (Ford 1998). At its construction, the home fronted the water, so that it was easy for travelers to disembark and for commercial interests to load and unload their wares. Site is again perhaps the most significant of the structure's architectural features.

Fig. 10-6 Carlyle House

The building itself, as mentioned, was influenced by Palladio's ideas and is perfectly symmetrical. Three stories tall, the rectangular building is of gray stone. A flight of stairs leads to the second story, the first story being half below ground level. The entryway and each end of the house are decorated with stone pilasters, and the entry door is surrounded by matching pilasters surmounted by a typical Roman arch. Simple rectangular windows range along all three stories of the façade, and two central chimneys adorn the roof. Two other homes in the Old Town area bear examination for their architectural power.

Robert E. Lee, the leader of the Confederate forces in the American Civil War, lived in this area while a boy. This boyhood home is an elegant townhouse done in the Federal style of architecture. The red brick building with white trim dates to 1795. Like Carlyle House, Lee's boyhood home has three stories and a small staircase with wrought iron rails leading to the second floor above a first story partially below street level. The entry façade is nearly flat but has a central portion that projects slightly from the rest of the wall at the top of which is a Classical entablature. The doorway is framed with pilasters beside a door topped with an arch. Rectangular windows range across the front of the house on both main floors and dormer windows adorn the roof. The final example of architecture in Old Town Alexandria to be examined is another residence of the numerous Lees.

Fig. 10-7 Lee Boyhood Home

The Greek Revival Lee-Fendall House was built in 1785 by Philip Richard Fendall, whose mother was a Lee. Between 1785 and 1903, thirty seven Lees lived in the home. In the middle of the 20th century, the building was home to John L. Lewis, the labor leader responsible for the establishment of the Congress of Industrial Organizations (CIO) and the President of the United Mine Workers (http://www.leefendallhouse.org). The home is white siding and has four elevations, the rooflines of which from the front look like stair steps. The tallest portion of the house is two full stories built over a story that is partially below ground and topped with a half story. A small circular staircase takes the visitor to the main entrance on the second floor, which is a small porch with a Greek entablature above the door. Rectangular windows with dark shutters range rhythmically along the front.

Fig. 10-8 Lee-Fendall House

In much the same way as political and commercial leaders did, wealthy landowners along the Mississippi River in the southern part of the US built impressive homes to demonstrate their wealth and status. Perhaps the best surviving examples of these are to be found in Louisiana in near proximity to the Mississippi River. These antebellum mansions include Oak Alley, Houmas House, and Shadows on the Teche. Southern Louisiana has an almost unique and exotic natural beauty, exemplified by placid bayous meandering through tall cypress trees and massive live oaks draped with Spanish moss and populated by plentiful wildlife. This natural beauty surrounds and complements the numerous mansions to be found there.

Perhaps the most striking such home is Oak Alley at Vacherie about halfway between Baton Rouge and New Orleans. The name comes from the entry road leading from the main road paralleling the Mississippi River to the house itself lined with ancient and huge live oaks that provide a canopy to the visitor arriving at the house. The oak lined pathway predates the house itself, having apparently been planted by a previous owner of the land, which means that the architect and builder of the present early 19th century structure recognized the dramatic visual potential of putting the building at the end of this road. Today, the typical photograph of Oak Alley is taken at a distance sufficient to show the "oak alley" leading to the mansion. The mansion itself is also visually dramatic. It is a two story structure built in the Greek Revival style. Symmetrical from all sides, there is a veranda around the ground floor and a gallery for the second floor. The home is decorated on all sides with 28 white Doric columns and three dormer windows on each side of the roof. A widow's walk in the center of the roof and brick chimneys top the building. The veranda extends outward from the house some 13 feet, which serves to shade the structure from the sun's heat during the day. The interior walls are sixteen inches thick for the same purpose, and the floors are 12 feet high (http://www.oakalleyplantation.com). Each floor has rectangular windows with shutters ranging rhythmically around the house.

Fig. 10-9 Oak Alley

Houmas House, also along the Mississippi River at Burnside near New Orleans, is almost a replica of Oak Alley insofar as its architectural style and visual impressiveness are concerned. There are many live oaks, some of which form an alley, but the dramatic effect of the Oak Alley trees is missing. The rear of the house connects to a smaller residence that predates the present building's completion in 1840. The Doric columns, the veranda, the gallery, the dormers, the widow's walk, and the windows are quite like those at Oak Alley. Although both houses are clearly Greek Revival in style, they also fit the architectural style of French Creole. Creoles were the children of Europeans born in the New World who developed their own adaptations of the conventions of their parents. Among the characteristics of French Creole architecture were "generous galleries," "a broad spreading roofline," and "multiple French doors" (http//www.nps.gov/history/nr/travel/Louisiana/architecture). This style integrated easily with the older Greek Revival style, which accommodated itself well to the heat and humidity of the Louisiana summer.

Fig. 10-10 Shadows on the Teche

Like Oak Alley and Houmas House, Shadows on the Teche was an antebellum home that was the center of a large sugar plantation and built between 1831 and 1834. This home was built along the Bayou Teche in New Iberia, west of the other two structures assessed here. Like them, however, it is of Greek Revival architecture with French Creole influences and could easily be a sibling of the other two. All three of these homes illustrate well the desire on the part of the residents or owners to demonstrate their wealth and influence in a visually dramatic way. They each depend on the central architectural elements of site, symmetry, ornament, and rhythm to produce their visual impact. Unlike the other residences considered here, however, these antebellum homes are located in one of the most visually stunning landscapes in the world. Add the striking color of the local azaleas, camellias, dogwoods, and magnolias when in bloom, and these residences are unique architectural statements.

Moving away from these grandiose residences, it is still possible to follow the theme of building homes to demonstrate wealth and status. As mentioned earlier, there is in virtually every city in America if not the world an area or areas which are notable for their residential architecture. Many modern suburbs feature gated communities for the extremely wealthy or reclusive, while most older cities focus the spotlight on their historic districts. Even in the author's small city of residence, College Station, Texas, promotes the prestige of its historic Oakwood subdivision. While College Station is a young city, with the historic district dating only to the first half of the 20th century, addresses in the area command a premium price on the real estate market.

Fig. 10-11 Oakwood Home

Site in College Station is so important that some recent residents have bought old homes only to raze them in favor of building impressive and expensive mansions. It is not surprising that one thing which contributes significantly to site is proximity to Texas A&M University's campus and its on-campus football stadium, the university's student body and alumni being avid football fans. Site is the most important architectural element in College Station as well as other cities, but impressive structures both old and new follow the patterns seen here for visual impact. There is perfectly ordinary and simple residential architecture but also examples of Federal, Georgian, and Greek Revival styles as well as variations on several other standard themes.

Fig. 10-12 The Painted Ladies

A number of miles southwest of College Station lies the city of San Antonio and its prestigious King William historic district. In the middle of the 19th century, this area was populated by many wealthy German immigrant families, who named one of its streets after King Wilhelm I of Prussia from which the district as a whole gets its name. Today, the district has become once again a prestigious address, and one which features several ornate architectural styles. Similarly, there are numerous wealthy and historically significant areas of San Francisco, California, which also feature visually compelling architectural styles. Pacific Heights is one, while Alamo Square is another. Alamo Square is in one of the hilly areas of the city and is an area where residents flocked to escape the fires and devastation of the earthquake of 1906. Along one side of the square, a row of houses known as the Painted Ladies rises up a slope like stair steps. All share a Victorian architectural style, tall and narrow three story mansions with peaked roofs and the delicate, if not frilly, curves of porches and columns. The site provides a dramatic view of the downtown skyline.

Finally in similar fashion, the hillside houses overlooking the bays in Wellington, New Zealand, offer residents an impressive view of the city's bay, and they are elevated above the lower levels of the city for the status that height provides. These villas in the older suburbs of Wellington "evolved during the late 19th and early 20th centuries from simple, flat-fronted single-storey colonial cottages with verandahs running across the front, to two-storey houses with projecting faceted bay windows. . . . Other features included the use of stained glass, large double-hung sash windows and balustrade porches or verandahs" (Buerger and Omar 2001). From an architectural point of view, the sites of these homes are their principle element of influence, although ornament and color are almost equally important.

Apartments and Hotels

In addition to homes, apartments are the principle places of residence for many of the world's population, especially in large cities. As indicated above, the types of apartments in the world vary from small spaces provided for people who cannot afford better residences to elegant and spacious luxury units. Some apartments are owned by the occupants, as is usually the case with homes, but most are leased from a landlord. Magnificent examples of luxury apartments include the Circus and the Royal Crescent in Bath, England, the many units to be found in the South Kensington area of London, and the Place des Vosges in Paris.

The Circus sits on a hill in the city of Bath situated around a circular plaza with trees at the confluence of Brock, Bennett, and Gay Streets. There are three matching structures of Georgian architecture designed in the 18th century by John Wood the Elder and completed by his son, who designed the Royal Crescent a short walk from the Circus along Brock Street. Each of the buildings is semicircular in shape, so that taken together they form a full circle. Each is three stories of beige Bath limestone fronted on each story with paired engaged columns. Three orders were used: Roman composite, which is similar to Doric, then Ionic and Corinthian. Between each of the paired columns are simple rectangular windows. Over the first story, there is a Doric entablature "with alternating triglyphs and 525 unique pictorial emblems in the metopes, including serpents, nautical symbols, devices representing the arts and sciences, and masonic symbols. The parapet is adorned with stone acorn finials" (http://blog.quintinlake.com/2010/4/26). From virtually any line of sight, the structures have a Palladian symmetry and remain a prestigious address in Bath even today.

Fig. 10-13 The Circus

Shortly after completion of the Circus, Wood the Younger designed and built the Royal Crescent down Brock Street. Still on an elevation, the Royal Crescent overlooks the largest green space in Bath, Royal Victoria Park. As its name suggests this structure of 30 townhouses forms a semicircle. It too is constructed of beige Bath limestone but differs from the Circus architecturally in several respects. The first floor is flat and unadorned save for entry doors separated by two rectangular windows. Engaged Ionic columns rise from the top of the first story to the top of the third floor, and the entablature and parapet above. Like the Circus, the Royal Crescent is a premier example of Palladian symmetry. Its elegant symmetry and Classic ornamentation distinguish it as a place of residence for those with the wealth to afford it.

Fig. 10-14 Royal Crescent

There are, of course, many elegant townhouses to be found in London as well, although few of them match those in Bath. However, the South Kensington area of the city is home to a number of quite visually distinctive residences and has come to be recognized as one of London's most prestigious addresses. Like townhouses all over the city, those in South Kensington are attached to each other and are vertical and comparatively narrow like the brownstones in Manhattan. In a typical row of such apartments in the area near the Royal Albert Hall, the structures are multi-storied with several entry porticos framed by Doric columns. The style is Georgian, and taken as a whole the structures have a Palladian symmetry. Wrought iron fencing protects entries below street level. There are balconies on the second floor and the windows are flanked by engaged columns. The elegance of this architectural style is further enhanced by the stark white color of the stonework.

Across the channel in Paris, a splendid example of apartment living may be found in the Place des Vosges in the Marais district of the city. Built between 1605 and 1612, the residences here flank each side of a perfect square (http://www.paris.com). The site is a principal architectural feature of the Place des Vosges and the square within is a green lawn planted with linden trees whose center features a large circular fountain. Broad pathways radiate away from the fountain in a star pattern. The 39 French Renaissance style houses which make up the apartment complex are of red brick with contrasting off-white stonework trim. A broad ground floor arcade with rhythmically repeated archways runs around the square.

Fig. 10-15 South Kensington Apartments

The peaked roofline has small dormers at regularly spaced intervals. The significant design elements which account for the visual power of the Place des Vosges are, as mentioned, site, then the solidity of the square arrangement and the resulting symmetry, and finally the colorful stonework. Over the centuries, many prominent French personalities lived in the Place des Vosges, including Victor Hugo whose apartment home is now a museum.

Although there are instances of people who live permanently in them, the normal use of hotels is as a home away from home. Like both homes and apartments, the variety of hotels is extreme. For the purposes of this chapter, however, two hotels will be considered which demonstrate the potential for compelling architecture. These are the Grand Hotel in Taipei, Taiwan, and the Adolphus Hotel in Dallas, Texas. Not all hotels attempt to house the visitor in extremely luxurious surroundings. The quotidian hotels/motels most frequently visited by business and vacation travelers merely provide clean and comfortable places to spend the night. The Holiday Inn chain, Ibis, Sofitel, and Microtel come immediately to mind. For some hotels, however, the goal seems to be to provide a temporary mini-palace for its visitors and an experience therefore that is indelible in memory.

Fig. 10-16 Place des Vosges

The Grand Hotel in Taipei is a fine example of Classical Chinese palatial architecture. The main building has twelve floors, and for many years the hotel was the tallest structure in Taipei. A double curved roofline adorns the traditional vermilion stonework building. The hotel opened in 1952, although the main building was not finished until 1973. The brainchild of General Chiang Kai-Shek, its purpose was twofold: to provide a place for visiting foreign dignitaries and to promulgate Chinese culture to the West. Those who have stayed there include US Presidents Nixon, Reagan, and Clinton, British Prime Minister Margaret Thatcher, and the author.

In terms of architectural elements used for visual influence, site is again perhaps the most notable. The hotel sits atop a hill above downtown Taipei. A broad circular driveway leads to the front gate, which is a typical Chinese passageway and brilliantly colored in vermilion, aqua, and gold, and then to the hotel itself. The interior is especially dramatic visually and illustrates the use of color as a principle architectural element. The lobby, for example, is a vast open space whose ceiling is supported by circular columns painted in vermilion. The carpet is red, and the ceiling a dark woodtone. Chandeliers adorn the ceiling as well. Although the rooms in the hotel are not all so visually dramatic, they are richly adorned with ornate dark wood trim.

Fig. 10-17 Grand Hotel Taipei

One other example from the US of a similarly extravagant hotel will suffice to conclude this chapter on residences, apartments, and hotels as examples of architecture as visually compelling forms. It is the Adolphus Hotel in Dallas, Texas. The hotel was deliberately conceived to be extravagant. The Dallas city fathers, seeking a hotel that would provide the city with status, asked the beer entrepreneur Adolphus Busch to construct the hotel in 1912. "Fame, fortune, and flamboyance came together in 21 stories of unabashed beaux arts splendor that critics have called 'the most beautiful building west of Venice'" (http://www.hoteladolphus.com). This statement is not hyperbole, as one who has stayed at the Adolphus and in Venice can attest. Queen Elizabeth II, Babe Ruth, Oscar de la Renta and other such notables have been guests of the hotel.

The lower floors have an exterior of off-white stonework featuring massive arched windows. The upper floors of red brick are less ornate and have simple rectangular windows and balconies. The topmost stories are also off-white and are ornately adorned with arched windows and spires. As with the Grand Hotel in Taipei, the interior of the Adolphus is its most ornate and extravagant aspect. It can only be compared to such structures as the White House or to the interior of a royal palace. Not only are the lobbies, the stairways, and the dining areas resplendent with dark wood walls and ceilings or gilded white ones displaying paintings in the palatial style, but also the rooms themselves are large and appointed with the finest furniture, bedding, drapes, and lighting.

Fig. 10-18 Grand Hotel Lobby

Architecturally, the Adolphus Hotel is situated prominently in the heart of downtown Dallas near the city's financial centers. Its early 20th century architecture has been outstripped by the more modern skyscrapers of the city, which will be examined in the following chapter, but the hotel has lost none of its palatial power. The Adolphus lacks the extravagant use of color characteristic of the Grand Hotel Taipei, but its rich omnipresent woods have a similar visual effect. Wherever one looks in the Adolphus Hotel, there is no lack of the visual power of architectural elements.

Fig. 10-19 Adolphus Hotel

This chapter and the previous one have concentrated on living spaces, while Chapter Eight examined "citadels of power." The following chapter returns to the theme of power with a focus not on governmental or political power but commercial power. Perhaps the most characteristic feature of the modern city, especially in the US, is its skyline, especially seen from a sufficient distance to comprehend it all, and in turn that skyline is defined principally by its skyscrapers, those towers of commercial power that some see as a metaphor for America itself. While they do not fit the rubric of buildings, there is another distinctive aspect of the modern American city which is essential to commerce and which is as powerful visually as the skyscrapers themselves. That is the bridges which cross the many waterways of the modern world.

CHAPTER ELEVEN

THE WORLD OF COMMERCE: SKYSCRAPERS AND BRIDGES

Although the idea of a skyscraper, a building which reaches breathtakingly upward to the heavens, is as old as the Tower of Babel story and the Ziggurat at Ur, today's tall buildings are a phenomenon of the late 19th and early 20th centuries. Skyscrapers are the product of at least two forces. The first is the expansion of commerce during the Gilded Age following the Industrial Revolution, especially in America's largest cities. Even today, Chicago, New York City, Houston, Dallas, Atlanta, and San Francisco remain largely devoted to commercial interests with a need for office space to house their businesses. The second force is a limited amount of land available for horizontal expansion, especially in cities like New York. The obvious solution was to reach for the sky, a choice made much more feasible with the advances in engineering and building materials of the late 19th century. Indeed, one might well ask whether or not skyscrapers represent an architectural achievement or merely an engineering one. It would be easy, as some have done, to dismiss such tall buildings as structures whose only salient features are, in Vitruvian terms, their strength and their utility.

It is not, as the previous chapters of this book have clearly indicated, a particularly common human trait to build anything essentially devoid of beauty, of compelling visual traits, unless absolutely constrained by money and materiel. So also, as it turns out, is the building of skyscrapers a case study in the desire to have visual influence. This view is shared by many writers, for all of whom there is a clear aesthetic dimension to the architecture of skyscrapers and for some a metaphysical one (Huxtable 1984, Schleier 1990, van Leeuwen 1988). Such a view of skyscrapers was not common at the beginning of their construction. In the early days, the view that they were both commercial and ugly was widespread in the writings of America's artistic set. "Shortly after the appearance of the skyscraper in New York and Chicago, its aesthetic viability became a hotly debated issue," and "it was not until the turn of the century that skyscrapers were finally accepted as subjects for the fine arts" (Schleier 1990, 6-7).

Huxtable (1984, 7) writes that "the skyscraper and the twentieth century are synonymous; the tall building is the landmark of our age" and also observes that "the skyscraper is still an art form. The tall building has that in common with all major works of architecture consciously conceived in aesthetic terms" (8). Van Leeuwen (1988) thinks that skyscrapers are not only art forms but also embody metaphysical dimensions. Dwelling on the Tower of Babel story, he likens the building of skyscrapers to efforts to reach heaven, calling the idea "the skyward trend of thought." Whether either architects or those who hire them consciously think in such ways or not, it may be a human characteristic to aspire to stand on heights above which no one can go and to build structures that require an upward look for contemplation.

Certainly also skyscrapers are about the demonstration of power and status. This is no less true than it was for government buildings and the domains of royalty. Size equates with power and status. Once a building has been certified as the tallest in the world, there is only a short interval before it is surpassed by another, as if owning or building or having the tallest of buildings establishes a business, an owner, a city, and a nation as the best in the world. For those who occupy such buildings, it is not the "skyward trend of thought" that is the key visual element but rather the outward and downward visual field. Tall buildings give their occupants a rich sense of power, both over other buildings and perhaps also their occupants.

There seems little reason not to examine skyscrapers as legitimate objects of compelling visual form. Bridges, on the other hand, pose a different problem. One must accept that the basic reason for a bridge, like that of a road, is to enable transit. Even more than skyscrapers, bridges also are engineering marvels rather than architectural ones. As such, strength and utility would seem to be their virtues. However, one must only consider two things to come to the conclusion that bridges are not merely functional structures. First, bridge design appears clearly to go beyond providing a means of crossing a chasm or a body of water. Builders seem to think of them also as things of beauty. Why else make the color, say, of the Golden Gate Bridge in San Francisco so important an ingredient? Why adorn a bridge with statuary, as in the case of the Charles Bridge in Prague, if not for visual appeal? Finally, why add towers to the Tower Bridge in London when, from a functional point of view, they could have been omitted?

The second thing to consider evolves from the first. Bridges and cloverleaf overpasses have become the objects of serious photographic art and, either as the cause or the effect, tourist destinations. Photographs of the Brooklyn Bridge and the Golden Gate Bridge may be found on postcards and on posters and also perhaps in exhibitions. Visitors to the south of France regularly go off the well-traveled paths to see the Pont du Gard and the incomplete Pont d'Avignon of children's song fame, and Venice visits would be hollow without a walk across the Rialto Bridge. So, as will be demonstrated, the world's bridges are visually compelling forms as well. Indeed, the world of commerce includes other visually interesting structures, such as train stations and airports that are worthy of assessment and which will be dealt with in passing in this chapter.

The Skyscrapers of Manhattan, Chicago, and San Francisco

Although the first skyscraper, according to most historians, was William LeBaron Jenney's 1884 Home Insurance Building in Chicago, and although the "city of Big Shoulders" according to the poet Carl Sandburg still boasts a skyline replete with tall buildings, the skyscrapers of Manhattan surely are primary in the popular imagination. It seems appropriate therefore to begin this assessment of the visual impact of skyscrapers with the Flatiron Building, the Chrysler Building, and the Empire State Building. These are only three of the multiple beautiful tall buildings in the city, but they are three which illustrate well the use of basic visual elements in architectural design.

Fig. 11-1 Flatiron Building

Indeed, the basic visual elements in assessing skyscrapers are perhaps fewer in number than those of other structures discussed in this book, and they may also be more uniformly shared. The most striking visual element of all, for example, is scale or size. Included within this element is the horizontal line which draws the eye upward. With this element of size is also the element of weight or mass, for skyscrapers are not only tall they have a sizeable footprint. Skyscrapers are also usually located on prime pieces of real estate, often anchoring other commercial developments and adjacent to similar commercial properties. In Manhattan, for example, the Empire State Building covers most of a city block and is near 30 Rockefeller Plaza, Madison Square Garden, and Grand Central Station.

Fig. 11-2 Chrysler Building

Were these to be the only basic visual elements used by architects of tall buildings, viewers might not place them in such high aesthetic esteem. So it is usually ornament that adds the final touch of beauty. Ornament in this case is often the result of creative uses of curves, angles, and geometric forms and not as in the case of cathedrals the addition of sculpture. In the examples which follow, ornament is frequently a very compelling feature.

The Flatiron Building, located where Broadway and Fifth Avenue join in Manhattan, is a prime example of the importance of ornament in the design of tall buildings. Flatiron is a nickname, with the real name being the Fuller Building.

The nickname comes from the wedge-shaped building's resemblance from above to a simple flat iron. The building was designed and built in 1902 by Daniel Burnham, whose work may be found in Chicago, San Francisco, and Washington, DC (Glancey 2006). At its completion, the Flatiron was the tallest building in the world. It was also quite ornate, especially in the upper and lower floors. There one finds the limestone structure decorated with columns, flowers, and urns. Although long superseded by subsequent taller buildings, the Flatiron remains today a nearly unique sight with its special shape and decoration.

Fig. 11-3 Empire State Building

Representing the familiar motive referenced above, the automotive mogul, Walter P. Chrysler, wanted to build the world's tallest building, which opened in 1930 (Glancey 2006). Unfortunately, the building was supplanted as the world's tallest in just the following year by the Empire State Building. However, the Chrysler Building, also in midtown Manhattan, may be for many viewers the most beautiful one in the city. Its architectural style is that of Art Deco, a product of the Paris Exhibition of 1925. "From the tip of its spire, 1,047 ft (319 m) above the Manhattan sidewalks, down to its Croesus-rich marble and chrome-steel lobby, it epitomizes the stylishness and happily decadent opulence of Art Deco" (437). That spire, of stainless steel, atop a silver gray stone structure appears, according to one writer, to be made up of "stylized Chrysler hub and radiator caps" (437).

As noted, the following year the Empire State Building was completed and remained the tallest building in the world for over 40 years. It remains the tallest in Manhattan and dominates the New York skyline. In popular culture, the building is familiar as the structure climbed by an escaped King Kong in the movie of the same name. In engineering annals, it is regarded as especially strong with its steel girder frame, strong enough to

Fig. 11-4 Willis (Sears) Tower

withstand the crash of a World War II-era B25 bomber. The building's exterior is, like the Chrysler Building, an Art Deco design, but it has nothing like the ornate spire of that building. What is distinctive and especially eye-catching about the Empire State Building is its stepped back construction designed to minimize the shadowing of other smaller buildings nearby. The structure is tiered like a wedding cake and resembles some of the Mesopotamian structures like the Ziggurat at Ur. There is a spire at the top of the building which serves as a radio tower and an observation deck that also figures prominently in such Hollywood films as *An Affair to Remember* and *Sleepless in Seattle*. In recent years, the building's upper stories have been lighted with successive floors of red, white, and blue lights, making the view of the Empire State Building particularly compelling at night.

Manhattan is known for its skyline, and it is no less true that Chicago is as well. Chicago's Madison Avenue, perhaps its most famous street, and the Magnificent Mile are described by the skyscrapers that rise along them and in the area of "the Loop" generally. Because it was also for a time the tallest building in the world, the Sears Tower, now the Willis Tower, is Chicago's most famous tall building. The tower is still the tallest building in the US at 110 stories and is a popular visitor attraction for its skydeck, which allows unobstructed views of the landscape for miles in all directions. Like the Empire State Building, which it supplanted as the world's tallest building in 1974, the Willis Tower's visual impact, aside from its height and size, comes from its stepped back construction. The effect of smaller and smaller tiers as the building rises keeps the structure from seeming merely monolithic as well as minimizing the shadows cast by the building.

Chicago's skyline is also marked by several other skyscrapers that are particularly compelling visually. The Trump International Hotel and

Tower is second only to the Willis Tower in height in Chicago and resembles it architecturally. One Museum Park is a condominium tower, rather than a commercial building, but is especially compelling visually for its stepped back rise, its gracious curves, and its generous use of glass. The Chicago Board of Trade Building, built in 1930, is one of the tallest Art Deco buildings in the world. Across the continent to the West Coast, San Francisco also boasts a remarkable skyline.

Fig. 11-5 Transamerica Building

San Francisco's tallest and most prominent skyscraper, and a building with unique visual appeal, is the Transamerica Building. The Transamerica Building's distinct visual appeal comes primarily from its shape. Rising from a four story rectangular base, it is a pyramid with a spire at its top. Wings jut out from the upper stories. Its size and shape are augmented by its dazzling white color, a function of the crushed quartz which covers its superstructure. Many urban areas in the US, as well as those in other countries, are defined by their downtown commercial centers and by the skyscrapers that populate them. Before assessing two distinctive European examples of such structures, two cities in Texas provide a final American collection of skyscrapers.

The Skyscrapers of Dallas and Fort Worth

Dallas and Fort Worth are the most populous two cities in north Texas, and along with the smaller cities which surround and connect them form a single large metropolitan area referred to as the Metroplex. As noted earlier in this book, the two cities have often been rivals in several areas and have traditionally been quite distinct culturally. Fort Worth, at least in the popular mind, was the "cowtown," while Dallas was its sophisticated and refined neighbor. Predictably, the two cities, as well as Houston in the south, have vied for the distinction of having the tallest buildings in the state. While Houston can boast the two tallest buildings in Texas, it is Dallas that has emerged as the most notable commercial center, and its skyline reflects this leadership. Indeed, the popular American television series, *Dallas*, opened with a view of the Dallas skyline seen from the south looking north. In truth, the skylines of both cities are especially notable in that they stand out on a high plain from the surrounding prairie.

Fig. 11-6 Fountain Place

Dallas's skyline includes some 30 tall buildings built between 1922 and 1985 which range in height from 29 stories to 72. Although a few of these structures are simply vertical rectangles, most were designed with a clear view toward visual impressiveness, arguably in a more distinct way than in Manhattan and Chicago. Certainly one of the most beautiful of these buildings is the blue-hued 6o story Fountain Place designed by I. M. Pei and Partners. Difficult to describe verbally, this building is a vertical rectangle whose upper stories change to a triangle.

Additionally, along one side of the structure is a projecting diagonal slash of additional space.

"The skyscraper is clad in green glass and is sculpted into a prism. One of the unique features is that it looks completely different as you view it from various locations in Downtown. Starting with a parallelogram in plan at the ground floor, upper stories are supported by two triangular shaped masses at the north and south corners. This configuration is present for the first five floors. Levels 6 through 13 are square in plan. At the 14th level, the area of wall between the face of the building and the angled base below starts sloping upward, allowing the parallelogram shaped tower to be visible" (http://www.dallas architecture.info/fountain.htm).

Fig. 11-7 Bank of America Plaza

The tallest skyscraper in Dallas at 72 stories, and the third tallest in Texas, is the Bank of America Plaza. Although modern in its architecture, the building features quite visually compelling design elements. "The plan features several notched corners which allow multiple corner offices per floor. Each of these notches stop at different heights at the top of the building, giving the obviously modern design an 'Art Deco' styling" (http://www.dallasarchitecture.info/ bankofa.htm). Renaissance Tower at 56 stories is the second tallest Dallas building and is visually distinctive for the double X pattern on its side which is now illuminated at night. In the *Dallas* television series, the Renaissance Tower was the home of Ewing Oil. It is topped by a small tower in the center of the flat roof.

Also quite visually distinctive are the 60 story Comerica Bank Tower and the 55 story JPMorgan Chase Tower. The Comerica Bank Tower was designed by Philip Johnson and John Burgee, New York architects, and features the ancient design feature of barrel vaults. It occupies an entire city block. "Its base is larger than the tower rising out of it and is capped off by a barrel vault. The tower starts off rectangular, and through a few setbacks forms a cross shaped tower. At the top each arm is capped by a barrel vault. The center sections of the building that ultimately become the vaults are glass curtain walls, while the remainder of the building is clad in granite" (http://www.dallasarchitecture.info/bankone.htm).

The JPMorgan Chase Tower is visually distinctive for its "keyhole" design in its upper reaches. To describe it, one writer says "its pink granite grid façade peters out to reveal a more delicate grid of glass and steel. . . . the gem of the building is truly at the top where the tower divides to reveal a . . . 'keyhole" as it is commonly known. . . . The eye is led from the base to the keyhole by a spine made from two protruding windows. Their antithesis is expressed through a ridge that dents the corners" (http://www.glasssteelandstone.com).

Fig. 11-8 JP Morgan Chase Tower

At the western edge of the Dallas downtown area stands a unique tall structure, although it could be compared to Seattle's Sky Needle and San Antonio's Hemisfair Tower. This is the 560 foot tall Reunion Tower, an observation tower built to be part of a Hyatt Regency Hotel. "The shaft of the tower is constructed from poured in place concrete and features four concrete cylinders. Three are placed on the outside in a triangular pattern and they house elevators. . . . The top of the tower is a three level structure, also constructed of poured in place concrete. Surrounding this top is a geodesic dome formed with aluminum struts." This dome contains the observation deck, a restaurant, and a cocktail bar which rotate (http://www.dallas architecture.info/reunion.htm). It is lighted at night by 260 lights.

Fig. 11-9 D. R. Horton Tower

Although neither so spread out nor so numerous as Dallas's skyscrapers, the tall buildings of Fort Worth are quite visually impressive in their own right. They center mainly in the reclaimed city center known as Sundance Square, which was previously described in Chapter Eight. Sundance Square itself is on a bluff in downtown Fort Worth, further accentuating the visual impact of all its buildings. Although by no means the only visually compelling tall building in Fort Worth, the D. R. Horton Tower in Sundance Square is a fine example of the skyscraper style. It rises 38 stories in the sky and stands out among the smaller restored buildings of the location. This building and a sister structure are in the shape of pinwheels. "The bases of both buildings are carved away creating covered areas beneath them. At the tops, each arm of the pinwheel stops at different levels, finally culminating with only the elevator penthouses left" (http://www.fortwortharchitecture.com/drhorton.htm). Fort Worth, Dallas, San Francisco, Chicago, and New York City are all home to some of the most visually compelling tall buildings in the world from which these cities derive great power and status as commercial centers. The same may be said also of other cities around the globe, although the skyscraper is a distinctly modern building innovation. However, even in some of the older cities of Great Britain and Europe, skyscrapers may be found in the more modern areas. Two particularly good examples are the so-called Pickle or Gherkin in London and La Défense in a western suburb of Paris. Both are unusual in their designs, even for skyscrapers, and are especially compelling visually.

The Pickle and La Défense

The Pickle is nicknamed as it is because of its resemblance to a gherkin standing on end. Its more mundane official name is 30 St. Mary Axe or the Swiss Re Building. Designed by Norman Foster and built where the Baltic Exchange Building was destroyed by a bomb in 1992, this tall building rises 41 stories into the London sky. "The building is stunning in appearance—a festival of dark and light glass with spandrels serving as streamers circling it like an Olympic dancer. Rectangular glass panes are forsaken for triangles and diamonds, and with each floor offset from the one below by five degrees, the building appears to work itself into a spiral like a pile of clay on a potter's wheel. The result is stunning and smooth. . . . the glass façade is actually all made of flat panels. The only curved piece of glass is the lens that tops the structure" (http://www.glasssteelandstone.com). The fact that this building stands in sharp contrast to the architecture of its older companions in this area near the famous Tower of London makes it all the more visually prominent.

Fig. 11-10 The Pickle

Because of the historic character of the Parisian city center, as well as the lack of space, tall commercial buildings in the area are to be found mainly in the western area of the city, across the Seine, called La Défense. The most visually spectacular of these skyscrapers is La Grande Arche de La Défense, a building that whose visual impact is enhanced by its special location. It stands at the western edge of an axis called the Triumphal Way that, moving east, includes the Arc de Triomphe de l'Étoile, the Champs Élysées, and the Louvre. On a clear day, one can see each of these monumental structures along the axis. "The project to build the 'Grande Arche' (Great Arch) was initiated by the French president Mitterand. He wanted a XXth century version of the Arc de Triomphe. . . . It is a 106 meters (348 ft) white building with the middle part left open. The sides of the cube contain offices" (http://www.aviewoncities.com). This account of a number of the world's most visually influential skyscrapers illustrates well the notion that even when one intends to construct a building whose main purpose is commercial there is virtually always a strong motivation to make the structure beautiful, to make it visually persuasive. While those other examples of commercial building, the world's great bridges, are perhaps even more likely to subordinate appearance to structural integrity, it may still be seen that the aesthetic dimension is not often ignored.

Fig. 11-11 La Grande Arche

Some of the World's Great Bridges

Fig. 11-12 Pont du Gard

The Romans were perhaps the world's most notable engineers. Wherever they went in the ancient world, they built whatever was needed to accommodate the needs of their military forces and those who accompanied them: baths, bridges, aqueducts, fortresses, and walls. However, the aspiration to aesthetic excellence seems always to have accompanied the Romans' desire to build. There is no better example of this aspiration than the Pont du Gard, which crosses the River Gard in southern France near Nîmes. Built in the 1st century BCE by Marcus Agrippa, this bridge was part of a 30 mile long aqueduct designed to bring water to the Roman city now called Nîmes (Cowan 2002). That it has stood so long is, of course, a tribute to the strength of Roman construction. There is no mortar, for example, holding this and other similar Roman structures together. That it is so visually compelling that it has remained a principal tourist attraction long after its usefulness as an aqueduct ended is a tribute to the Roman quest for beauty.

Fig. 11-13 Bridge of Sighs

The Romans were the first to use the stone arch as a principal architectural feature, and archways have graced buildings of all kinds through the ages as has been seen in previous chapters of this book. They provide the eye with the gracefulness of vertical lines topped with curves as well as the rhythmic repetition of patterns also so common in Western architecture. The Pont du Gard is "a stone bridge of three tiers of arches. . . . The lowest level of the Pont du Gard consists of six arches, 67 feet (20.5 m) high, the longest spanning 80 feet (24.4 m). The middle level has six arches, 64 feet (19.5 m) high. The upper level has 35 arches, 24 feet (7.4 m) high; the covered water conduit on top was a deep cement-lined channel" (Cowan 2002, 188-189).

Like the Pont du Gard, the Bridge of Sighs in Venice was an almost purely utilitarian structure. It was constructed to allow a passageway from the Doge's Palace to dungeons across one of the city's many canals, and its purpose was grim. It provided a pathway for the condemned on their way to their cells who were thought to "sigh" at their last sight of freedom. Yet, as was the case throughout Venice, the bridge was beautifully adorned, so that its exterior at least would draw the eyes of passersby. The short bridge is made of white limestone and has two windows. The upper portion curves attractively and has sculpted scrolls on top. Below the windows, there is another curve with sculpture at each corner.

The Rialto Bridge, which resembles the Bridge of Sighs in appearance if not in size, is one of the major pedestrian passageways linking important land areas together in the city of canals. It crosses the Grand Canal, which is Venice's primary waterway. The present bridge dates to 1591, although there were previous bridges in this location. The current bridge is of masonry with timber supports (Cowan 2002). A central arch rises 21 feet above the water so as not to impede water traffic. A railed banister lines each side of the base, and the superstructure is a simple walkway flanked by rows of shops on each side. These shops are formed by a successive series of arched passages with a taller central arch at the peak of the bridge. Like the arches of the Pont du Gard, the Rialto Bridge's arches provide the rhythmic repetition and symmetry common to Italian Renaissance architecture.

Fig. 11-14 Rialto Bridge

Classic examples of visually compelling spans dot the European and British landscapes, with the Tower Bridge in London, the Charles Bridge in Prague, and the Chain Bridge in Budapest being among the most impressive. The Tower Bridge, not to be confused with the London Bridge of children's rhymes fame and which no longer exists in London, crosses the River Thames near the site of the Tower of London. Its chief visual characteristics are its two Gothic stone towers, complete with spires on each of their four corners, and the pedestrian walkway which connects the towers near their tops. Beneath the walkway is the roadway which is a drawbridge allowing the passage of tall ships. On the land end of each tower, the suspension supports for the bridge curve gracefully downward. In spite of its use of medieval design elements, the bridge is comparatively new, having been constructed between 1886 and 1894. Some have criticized this use of earlier architectural elements as having been too extravagant (Cowan 2002), but their use is both visually compelling and fits well in the immediate geographic area.

Fig. 11-15 Tower Bridge

With the collapse of the Soviet Union and the emergent democracy in the Czech Republic, Prague has become a major European tourist destination. As noted in a previous chapter, Prague has many striking architectural elements, not the least of which is the sculpture lined, now pedestrian Charles Bridge over the Vltava River. This bridge links the Prague Castle area of the city with its commercial and residential center.

Fig. 11-16 Charles Bridge

The construction is of sandstone with a base of 16 Roman style arches. 30 statues along with lampposts line either side of the bridge, which ends on the Old City side in a tall Gothic tower. These simple architectural elements underlie the beauty of the Charles Bridge and undoubtedly contribute to its popularity with tourists and residents alike. There is hardly a time of the day when the bridge is not crowded with strollers.

The Chain Bridge in Budapest links the Buda side of the Danube Riber with the Pest side near Buda Castle on that side and the Hungarian Parliament on the other. Two stone towers stand as the points of anchor near each side of the river with the steel suspension strands gracefully curving between. Stone lions guard the base of the towers on each side. Although the bridge is primarily a road bed, there are pedestrian walkways flanking it on each side. One of the most striking visual elements of the Chain Bridge is the resemblance of the two stone towers to traditional Roman triumphal arches.

Fig. 11-17 Chain Bridge

Like Europe and Britain, the US has numerous bridges whose striking visual appearance has made them familiar to souvenir hunters and film goers. Perhaps the two most familiar are the Brooklyn Bridge, connecting the New York City boroughs of Manhattan and Brooklyn, and the Golden

Gate Bridge, connecting the city of San Francisco with the more rural Marin County areas of Sausalito and the Muir Woods. The Brooklyn Bridge is similar in construction to the Chain Bridge in that it is a steel suspension structure anchored on each side by stone towers. The Brooklyn Bridge towers have two peaked arches, one for each roadway direction, with a pedestrian walkway between.

Fig. 11-18 Brooklyn Bridge

For many, the most visually dramatic bridge in the US is the Golden Gate Bridge in San Francisco. In most ways, this bridge is not different from either the Chain or the Brooklyn Bridges. It is anchored on each side by stone towers and has a conventional steel suspension structure which swoops gracefully from tower to tower. The two towers are dramatically taller than those of the other bridges however, and they consist of four arched passageways stacked one upon the other. The length of the span between the two towers is also quite long, which enhances the soaring effect. Unlike other bridges, the Golden Gate is painted international orange, which gives it a visual distinctiveness unique in the world. As in the case of other notable structures, the site of this bridge contributes as much to its mystique as anything. Though it crosses San Francisco Bay, one side of the bridge is the open Pacific Ocean, which offers a vast, blue counterpoint to the bridge's orange paint.

Fig. 11-19 Golden Gate Bridge

Fig. 11-20 Sydney Harbor Bridge

The final bridge to be examined in this brief account of visually prominent structures is the Harbor Bridge in Sydney, Australia. Of all the bridges considered here, this one is maybe the least visually attractive. Its appearance has led locals to liken it to a giant coat hanger. It has double concrete pylons or towers at each end, but it is of steel through arch construction rather than suspension cable. Its visual appearance is quite compelling though, with size and mass being the essential elements. Its location on Sydney Harbor is also an important feature of its impact, and the addition of the Sydney Opera House across the inner harbor complements the stark visual power of the bridge.

Fig. 11-21 Victoria Station

The impulse to visual impressiveness in the commercial world extends beyond skyscrapers and bridges of course. Travelers not only use the bus and train stations of the Western world and the airports and subways as well, they flock to them for their compelling form. Victoria Station and St. Pancras Station in London are but two such examples, although they differ substantially in their appearance. Nineteenth century Victoria Station is comprised of three buildings, the original of the three being done in Edwardian style and adorned by sculpture. St. Pancras Station, also a product of the late 19th century, is red brick Gothic Revival. Although it is a world away from London, the British legacy of imposing railway stations can still be seen in the early 20th century Flinders Street Station in

Melbourne, Australia. Done in a massive Baroque style, the station's red and cream exterior is an icon of the city.

Fig. 11-22 Flinders Street Station

Although the Gare d'Orsay in the middle of Paris has been converted into an art museum, it was described at its opening as a work of art for its Beaux Arts elegance and its design intended to complement other buildings in its area. Its site is along the River Seine across from the Tuileries Gardens and the Louvre, a location which contributes directly to the station's visual impressiveness. Finally while considering Paris in terms of commerce, no one can disregard the many Art Deco entrances to the Metro scattered over the City of Light or the many underground stations which are beautifully decorated with paintings and other artwork.

There is also a nice coterie of visually impressive airports to be studied for their visual impact, although their locations and size make it difficult to survey them as things of beauty. However, who can ignore the visual impression of the TWA Terminal, designed by Eero Saarinen, at John F. Kennedy International Airport on Long Island in the US, or the sweeping grandeur of the main terminal at John Foster Dulles International Airport in Chantilly, VA, outside Washington, DC, also designed by Eero Saarinen? As frequent air travelers know, it is a visual delight to arrive in Amsterdam's Schiphol Airport, or Singapore's Changi, and Paris's Charles de Gaulle. It is quite surely not for nothing that Roth (1993, 1) calls architecture “the unavoidable art,” which it surely is even in the utilitarian world of commerce.

CHAPTER TWELVE

MONUMENTS AND MEMORIALS: THE SYNTHESIS OF ART AND ARCHITECTURE

J. DONALD RAGSDALE
AND FRANCES E. BRANDAU-BROWN

Just as bridges raised a question of appropriateness in the preceding one, so in this chapter one must answer the challenge that monuments and memorials are not architecture at all but are rather art. Of course, the author's perspective in this book and earlier ones (Ragsdale 2007, 2009a, 2009b) is that architecture is itself an art. However, it is surely true that such structures as the Eiffel Tower, the Washington Monument, and the Vietnam Veterans Memorial are not buildings *per se.* Such other structures as the Prince Albert Memorial, the Lincoln Memorial, and the Sir Walter Scott Memorial feature a statue of the memorialized subject within an enclosure resembling a building, while others place a figure atop a spire as in the case of the Siegessäule and Nelson's Column. Finally, such memorials as the Korean War Veterans Memorial and the FDR Memorial in Washington, DC, are tableaus, although they are accompanied by walls that partially enclose them.

It would be easy enough to classify monuments and memorials as art and exclude them from these assessments. Hanna (2002, p. 218), for example, takes this position. She writes that "monuments and memorials are not designed as private shelters or as public buildings for worship, administration, or entertainment, but to enshrine communal memories. They are art rather than architecture." However, taking this position ignores some salient realities. Most monuments—and from this point forward this term will be used inclusively for both structures—if not all have been designed by architects and not commissioned to sculptors or other artists. Second, monuments characteristically utilize the elements and principles of architectural design noted in Chapter One of this book and applied throughout the remaining chapters. Moreover in anthologies of architecture (e.g., Field 2007, Glancey 2006, Strickland 2001), monuments are included alongside the usual palaces, cathedrals, and chateaux. Especially insofar as the second reality is concerned, it would seem to be most productive to assess monuments as types of architecture along with the other structures in this book. Truth be told, monuments are a synthesis of art and architecture and in many cases feature elements of both.

In this chapter, monuments will be considered primarily from the standpoint of architecture. To facilitate the task of evaluating a hybrid phenomenon, a typology of monuments is a useful beginning point. Monuments are legion and they are ubiquitous, such that those assessed as visual persuasion here are only perhaps the most prominent ones in the Western world. A typology of these, therefore, should not be thought of as either exhaustive or definitive. Nonetheless, there are four clear types of monuments to be assessed below. The first type is the familiar statue of the memorialized person placed within a building-like structure. Such monuments are also commonly called shrines. Examples include the Lincoln and the Jefferson Memorials in Washington, DC, as well as the Chiang Kai-Shek Memorial in Taipei, The Republic of China. Stretching the designation somewhat, similar structures with the figure outside will be included. The second type is the pure geometric form, such as the Vietnam Veterans Memorial in Washington, DC, and the Arizona Memorial in Pearl Harbor, Oahu, Hawai'i. The third class of monument is the one least like other types of architecture. It is the tableau, such as can be seen in the Korean War Veterans Memorial and the Franklin Delano Roosevelt Memorial in Washington, DC. Finally for the purposes of this treatment there is the image on a spire. The monuments of this type include the Nelson Column in Trafalgar Square, London, and the Siegessäule in Berlin.

The Statue within a Building or Shrine

The Lincoln Memorial in Washington, DC, if not the most famous monument in the world, must certainly be the most familiar, and it honors perhaps America's most beloved president. It is appropriate, therefore, that it should have been designed in the fashion of a Classic Greek temple. It is a rectangular structure of marble with 36 fluted Doric columns ranging around the building. The number of states constituting the Union at Lincoln's death was 36, and the frieze running along the top of the columns names these states and includes the dates of their admission to the Union. All of the visually compelling architectural features of such temples as the Parthenon in Athens are present. The 44 foot high columns provide a symmetrical façade and a powerful rhythmic repetition. The white marble provides an image of elegance and purity. While not built on a promontory, the Lincoln Memorial is situated on the National Mall at the end of a long, rectangular reflecting pool, which points toward it. The sight lines are such that the monument is visible in every direction, unobscured by any other structure.

Fig. 12-1 Lincoln Memorial

One ascends a broad staircase to enter the building, where a statue by Daniel Chester French of Lincoln seated dominates the view. Inscribed in the interior walls are passages from the Gettysburg Address and the Second Inaugural Address, both of which speeches have been anthologized in American textbooks as singular examples not only of Lincoln's rhetorical ability but of great American speeches in general. The statue, in spite of the figure of Lincoln being seated, is visually powerful. This power derives from the statue's size and clear lines and from the suggestion of the assassinated President as one waiting for supplicants.

Within sight of the Lincoln Memorial is America's other statue within a building honoring a president. This is the Jefferson Memorial, which sits on the Potomac River Tidal Basin. This building shares some architectural features with the Lincoln Memorial, but it most closely resembles the Roman Pantheon. The Jefferson Memorial features Ionic columns, but the interior of the building is open to the elements. The building is circular and domed and has marble steps leading up to a portico with a typical Greek pediment. Within, there is a standing bronze statue by Rudulph Evans of the nation's third president and the author of the Declaration of Independence. Passages from the Declaration and other writings of Jefferson are inscribed on the walls. Like the Lincoln Memorial, the Jefferson is visually compelling for its rhythmically repeated columns, its radiant white color, and its traditional Greek design.

Fig. 12-2 Jefferson Memorial

Outside the US, there are a number of excellent examples of the shrine. Perhaps the most ornate of these is the Prince Albert Memorial in London's Hyde Park across the street from and facing the Royal Albert Hall. The enclosure itself is a simple square, open on all sides, but done in the most extravagant Gothic revival style possible. Atop the four corner posts rest triangular pediments and, at their intersection, a tall spire with a cross rises. Smaller spires rise at each corner. Each tympanum features a colorful painting. Within is a seated gilded bronze figure of Prince Albert. The shrine itself rests on a stone base on each corner of which are white sculptures representative of the four corners of the earth. The site is elevated from street level slightly and nothing nearby obscures the view. Many visitors may view the monument as gaudy, and the Gothic style as somewhat out of place. However, the richness of the ornamentation is homage to Victoria's love for her husband, and the Gothic style recalls some of the most moving and compelling architectural features ever developed.

Fig. 12-3 Albert Memorial

Victoria herself fares at least as well as Albert in the monument to this queen sitting outside the main entrance gates to Buckingham Palace nearby. There are many statues which serve as monuments to Queen Victoria, including one in the center of the city of Sydney, Australia, but there would appear to be none so ornate or so large as the one at the residence of the British monarch. The monument to the queen is not a building or structure within which sits a statue, like the previous examples here, rather it is a square white marble structure with the statue of the queen on the side facing away from the palace. However, it is clearly a shrine. On the west side, opposite the queen's sitting sculpture, is a statue of Charity. On the north is the Angel of Justice and on the south side the Angel of Truth. A bronze statue of Victory tops the monument. The Victoria Memorial, less ornate than the memorial to Albert, is nonetheless visually compelling. Its white marble and bronze ornaments shine brightly in the sun, and its location outside the palace is a site unmatched within the city of London for its significance. The statue also stands at the head of the street along which the royal guards march to and from the palace for the celebrated changing of the guard.

Fig. 12-4 Queen Victoria Monument

Fig. 12-5 Scott Memorial

Although not a royal memorial, the Sir Walter Scott Memorial in downtown Edinburgh, commemorates Scotland's best-known and perhaps most beloved author. The design resembles in some respects that of the Albert Memorial, although it is larger, has an even more massive central spire, and is more replete with the twists and turns of the Gothic style. The monument is of sandstone, which unfortunately has attracted over the years a black coating of soot from the chimneys of the city. Within, however, the statue of a seated Sir Walter Scott and his dog Maida is done in white Carrara marble. Although the site of the statue is in the Edinburgh city center, it is located in a park area set off from the more modern buildings of the city. Like other monuments of its type, the Sir Walter Scott Memorial gets most of its architectural impressiveness from its ornate style and its great size. In like manner, the National Monument to Vittorio Emanuele II in Rome is impressive for its style and size, although it looks nothing like any of the preceding monuments.

Fig. 12-6 Vittorio Emanuele II Monument

This monument to Vittorio Emanuele II is situated such that one cannot avoid the symbolism of the past, for it is located on a hill at one end of what was the Roman Forum. It also commands a powerful view from the other side, which is a crowded avenue at the heart of the old city. Not only is the monument massive, it includes several features not found on other monuments. The monument has a strong horizontal line. Broad but shallow steps lead up to a platform on which is erected a circular pedestal atop a semicircular base. Both base and pedestal are of marble and feature carved scenes memorializing episodes significant in recent Italian history.

Atop the pedestal is a massive bronze statue of Vittorio Emanuele II, who was the first king of Italy after its unification. Flanking the stairs at street level are four marble pedestals with bronze sculptures. Rising above and behind the statue of the king is a building which appears to be one long portico featuring Corinthian columns. At each end of this portico are projections in the form of Classical temples and with Greek pediments. Atop each of these porticos are bronze *quadrigas,* which were chariots drawn by four horses and often driven by Victory. The last example of this monument type is possibly the largest of all. It is the Chiang Kai-Shek Memorial Hall honoring the first President of the Republic of China, commonly known as Taiwan. The monument is in the capital of Taipei.

Fig. 12-7 Chiang Kai-Shek Memorial

A huge park in downtown Taipei is the setting for the Chiang Kai-Shek Memorial Hall. It is located on the eastern side of Liberty Square and is flanked toward the west by the National Theatre and the National Concert Hall. Most of the architecture in downtown Taipei is quite modern and has a distinct Western tone. The buildings in the park, however, are of traditional Chinese styling and resemble Buddhist temples. The National Theatre and National Concert Hall are decorated in rich tones of red and bronze, but the monument to Chiang Kai-Shek is a square of white marble whose sides slope gently toward the octagonal roof of blue tiles. The roof is actually a double one, with one atop the other. The white and blue colors of the monument, plus the red flowers in the gardens nearby depict the colors of the flag of the Republic of China.

A double staircase leads up to the entrance to the monument. Each has 89 steps, which number corresponds to the leader's age at his death. Within the monument, a bronze statue of Chiang Kai-Shek sits on the wall opposite the entrance. The interior walls are off-white, grayish stonework with a carved wooden ceiling and a variegated tile floor. Architecturally, the elements which make this monument compelling are site, mass, color, and balance or symmetry. The design is simple and unadorned providing a sense of quiet elegance. That it stands at an elevation which provides a fine panoramic view of Taipei also contributes to the monument's visual impressiveness.

The Pure Geometric Form

Fig. 12-8 Washington Monument

A number of the world's most famous monuments do not include the depiction of a person and often memorialize an event rather than an individual. The monuments of this type assessed here can best be described as pure geometric forms. Some, such as the Washington Monument, are simple obelisks. While not an obelisk, the Eiffel Tower is a similar spire. Others are tablets laid against an earthen base, such as the Vietnam Veterans Memorial, and some, like the Arc de Triomphe, are massive arches reminiscent of similar victory arches of the ancient world. Finally, there are what appear to be merely random blocks covering a field like a cemetery, such as Berlin's Memorial to the Murdered Jews of Europe. All of these types of monuments derive their visual force from the striking lines used by the designers to draw the viewer's attention.

Standing at the opposite end of the National Mall in Washington, DC, from the US Capitol and anchoring this city of monuments like a majestic road sign, the Washington Monument is a simple, but massive, obelisk. It gains its visual significance from its site and from its size. Somewhat like the Arc de Triomphe in Paris, from which radiate so many streets that the view from above is of a great star—called *l'étoile* by the French—the

Washington Monument is easily the most commanding presence on the Mall. It rises a bit over 555 feet in the air. The monument of marble slabs slopes gently upward and is topped by a pyramid. Its vertical lines are a fitting tribute to the nation's first president.

Fig. 12-9 Vietnam Veterans Memorial

Looking toward the Lincoln Memorial from the Washington Monument, there is to the right of the Reflecting Pool a monument to Vietnam Veterans of unique minimalist design by Maya Lin. The design was not well received by many, including many Vietnam veterans, because of its spare qualities and an alternative memorial to Vietnam veterans, a traditional bronze statue of three comrades in arms, stands nearby. The monument is composed of two slabs of gabbro polished to a high gloss which contain the names of all those killed in the war in order of their deaths. One section of the wall points toward the Lincoln Memorial, while the adjoining one points toward the Washington Monument. The slabs are set against the earth and grass behind them, and a walkway takes the visitor gently downward and then back up, since the height of the walls, or the depth of the walkway, varies in order to accommodate the greater number of names at the height of the conflict.

The wall, as it is simply called, may well seem too minimal for some. However, it has great visual power. Its strong horizontal line is complemented by the depth at the point of the two slabs intersection, and the consequent symbolism of the depth of the conflict and the number of deaths cannot be missed. The dark color of the slabs is appropriate for this most troubling of America's wars. The wall's site, too, contributes to its visual power, since it stands adjacent to the Reflecting Pool and within the shadow of the Lincoln and Washington monuments. It now has perhaps even greater significance since the completion in 2004 of the US National World War II Memorial between the Reflecting Pool and the Washington Monument.

Fig. 12-10 World War II Memorial

The US National World War II Memorial is impressive for its site if for nothing else. Located at the end of the Reflecting Pool opposite the Lincoln Memorial, its only detracting feature is that it adds to the congestion of so many monuments placed close together. However, it is a great open structure that does not block the view of any other monument. The WW II monument is a semicircular ring of granite tablets, each of which is 17 feet high, around a large oval pond and fountain. Each tablet is inscribed with the name of one of the 48 states of the Union at the time of the war and that of other US territories such as the US Virgin Islands. The rings, as noted, do not completely encircle the pond, and each has as its centerpiece an arched tower open on all four sides. One tower is identified as "Atlantic" and the other as "Pacific," representing the two theatres of battle in which the US was involved. Each tower is ornamented with sculpture. Also around the pond at the base of the tablets are shorter tablets depicting scenes from the battles of the war. The monument is, like so many others, stark white. Architecturally, the rings of tablets establish a rhythmic repetition of vertical lines and an airiness that is necessary in such a large monument to guard against oppressiveness.

Fig. 12-11 Arizona Memorial

Far away to the southwest, an even more poignant WW II monument sits in Pearl Harbor, in Honolulu, on Oahu in Hawai'i. This is the stark building which sits on pillars above the hulk of the battleship USS Arizona, sunk by Japanese aircraft in the attack on December 7, 1941, which brought the US into WW II. Not only is this site visually compelling for being the place where the earliest casualties of the war took place, it allows the visitor to see the sunken ship and note that oil still seeps from its hold and that some of the ship's superstructure still stands above the water line. The experience is quite humbling, and few visitors make any other sound than a whisper. The building which is the monument itself is a stark white bridge-like structure with seven openings on each side and seven in the ceiling, each representing the day of the attack and together representing a 21 gun salute. Outside, the entrance wall and the one opposite it slope gently to two peaks which are separated by a downward curving ceiling analogous to the suspension cables of a bridge. The structure is reached only by boat. At the end of the interior opposite the entrance is a wall inscribed with the names of those on the Arizona who perished on the day of the attack. While it is the location of this monument that is most compelling visually, it is also the stark white color and the simplicity of the structure which contribute to its power.

Fig. 12-12 Eiffel Tower

A monument of a much different type than the ones discussed earlier and one which was never intended to memorialize anything or even to remain standing beyond its initial purpose to mark the entrance to the World's Fair of 1889 has become the foremost icon of the city of Paris. This is the Eiffel Tower. Although rather hard to describe, the Eiffel Tower is easily one of the most recognizable structures in the world and one of the most visited in Paris. It was designed so that it could be readily disassembled, but it was useful as a location for radio signals and kept. It is taller than any other structure in Paris and may be the city's most iconic piece of architecture. Critics of the tower at the time of its construction accused the open iron structure of ugliness. Certainly, in a city of beautiful stonework, sculpture, and Baroque architecture, it is easy to see how the tower might have impressed early viewers. From the standpoint of the architectural elements used in its design, however, it is quite visually compelling.

There are many "ideal" sites for buildings in Paris, but surely the location of the Eiffel Tower must be included among them. The tower stands on the right bank of the Seine River and overlooks the Palais de Chaillot. On the east side of the tower is the Champs de Mars and the École Militaire. This location in the Seventh Arrondissement is also a very wealthy one. Time has resulted in a greater tolerance for open structures of metal, but surely there is more to the tower than merely its component parts. The tower stands on four massive legs and rises like an obelisk to the sky. The base of the structure is more massive than it appears in photographs and gives the impression of powerful size. Up close, as well as from a distance, the exposed ironwork has an almost lace-like quality of great visual attractiveness, and the vertical expanse not only draws the viewer's eyes upward but dominates the Paris skyline. In a similar way, the Arc de Triomphe across town dominates the view of arguably the most famous avenue in the world.

Vying with the Eiffel Tower for the designation of Paris's most iconic structure, the Arc de Triomphe, which stands at the western end of the Champs Élysées, is a monument in the Classic tradition of triumphal arches from Egypt, Greece, and Rome. It was commissioned by Napoléon I to commemorate the military successes of the Grand Armée. Although not completed in Bonaparte's lifetime, the Arc de Triomphe is the largest and best known example of this fundamental architectural form. It bears numerous bas-relief sculptures on the sides of its square supports representing the images of war and also now features the eternal flame of the tomb of the French unknown soldier. Its location is a spectacular site. It is the Place de l'Étoile, the star, and the confluence of twelve avenues radiating from the circular location. There is also a modest elevation of this place from the surrounding neighborhoods. The stonework is massive even without the sculptures, thus the visual power of the monument is largely a function of scale and mass. Another war memorial of a sort may be found in Berlin, capitol of the newly reunited Germany but also the capitol of Nazi Germany. This is the Memorial to the Murdered Jews of Europe during World War II.

This monument is the last to be considered of the pure geometric form type, but it is also a tableau of sorts such as the ones to be discussed in the subsequent section of this chapter. This monument covers a 5.5 acre site in

Fig. 12-13 Arc de Triomphe

the shadow of the Reichstag a block south of the Brandenburg Gate (http://www.nytimes.com/2005/05/09/arts). Again, the site is an integral part of the structure's architectural power. It is near the site of Hitler's chancellery and his underground bunker. However, it is the nature of the monument itself that is most visually impressive and at the same time most disquieting. There are no signs to identify the structure. What one sees is a field of 2,711 pillars of varying heights, which suggest tombstones or mausoleums. Clearly, the intent was to evoke the grimness of death, but there is enough ambiguity in the geometry of the pillars that one feels pulled by the sheer mass and scale of the site. The passageways between the pillars, and visitors readily take them, suggest a cemetery. While not beautiful in any conventional sense, the stark visual power of this memorial is palpable. While the Memorial to the Murdered Jews of Europe is abstract in its design and therefore not truly a tableau, there is no mistaking the pull of the populated space.

Fig. 12-14 Memorial to the Murdered Jews of Europe

The Tableau

The idea of a tableau is that of multiple representations of people and events. Tableaus adorn many temples, museums, and government buildings, but the monuments to be discussed here are somewhat unique freestanding tableaus. They are recent additions to the multiple monuments in Washington, DC, and represent belated efforts to honor first an event and then a US President previously slighted in the nation's history. The first is the Korean War Veterans Memorial, and the second is the Franklin Delano Roosevelt Memorial. Occurring as it did just between the more momentous World War II and the more protracted and unsuccessful Vietnam War, the Korean War is one of America's two largely forgotten wars. World War I is the other, and efforts to memorialize it are underway presently. Franklin Delano Roosevelt was America's longest serving president and the architect of the domestic policies of the New Deal and the foreign policies of World War II. Both monuments are located in the same general area of those previously discussed. The Korean War Veterans Memorial is located across the Lincoln Memorial Reflecting Pool from the Vietnam Veterans Memorial, while the FDR Memorial sits on the Tidal Basin near the Jefferson Memorial and amidst Washington's famous Japanese cherry trees.

The Korean War Veterans Memorial is a visually arresting collection of 19 stainless steel statues of a squad of soldiers walking up a gently sloping hill. They are spread out in the form of a triangle. The troops depicted are in full battle dress, and their display on a grassy field imparts a strong sense of reality. The sculptures also represent the diverse backgrounds of the soldiers from Caucasian, to African-American and Hispanic, to American Indian.

Fig. 12-15 Korean War Veterans Memorial

Arranged around parts of this triangular tableau are an Honor Roll, a Pool of Remembrance, and a Mural Wall. The latter is a polished black granite wall, similar to the Vietnam Wall, with 2,400 images etched upon it based on 15,000 photographs from the Korean War. The images are of every sort of participant in the war, from pilots, gunners, corpsmen, chaplains, engineers, and the like from every branch of military service. The visual power of the Korean War Veterans Memorial is surely due in part to its site, but the array of soldiers in the field is the most striking feature of the monument. The horizontal lines of the adjacent walls frame the statues and provide texture, and the black granite Mural Wall gives the monument its stark color. The stainless steel statues glow in sunlight.

Fig. 12-16 FDR Memorial

The FDR Memorial is a series of statues representing events in the presidency of this man and of FDR himself. The president himself is represented seated in his wheelchair, a depiction of which FDR tried assiduously to avoid in photographs during his life so as not to appear weak. He had, of course, been partially paralyzed by poliomyelitis. Since FDR was elected to four terms, the design of the monument is suggestive of four rooms. There are no real rooms, as such, but rather partial walls suggestive of enclosed spaces. Each room represents an event of significance in FDR's presidency: Breadline, Rural Couple, the New Deal, World War II. Interspersed throughout the monument are a number of waterfalls, which adds a natural texture to the site. Beside one of these is another statue of FDR seated next to his dog Fala. Finally, there is a standing statue of Eleanor Roosevelt, the only statue of a First Lady in a Washington, DC, monument.

The FDR Memorial spreads out over 7.5 acres of Tidal Basin frontage, and both the location and the size of the structure contribute directly to its visual power. Size and the ornament resulting from the varied sculptures also make major contributions. When the cherry blossoms are in bloom in the Spring, the FDR Memorial is especially colorful. Having examined the statue within a building, the pure geometric form, and the tableau as types of monuments, it is time to consider the last type, the statue atop a spire, as a means of visual persuasion. Examples of this type of monument range in time from antiquity to the 20th century and in geography from Rome to the New York harbor.

The Statue Atop a Spire

Fig. 12-17 Trajan's Column

Visitors to Rome have no dearth of sights which represent not only the glory of the Eternal City but also the architectural contributions of Roman builders. These include the arched entryways of the Colosseum, the triumphal arches in the Roman Forum, and the Baths of Caracalla, to mention but three, and also the often imitated statue atop a spire. Perhaps the best known of these monumental types is Trajan's Column adjacent to the Forum. The column was erected to commemorate Trajan's final victory over the Dacians. At its top, a statue of Trajan was erected, although that statue was eventually removed and one of St. Peter was put in its place. What a spire does was quite clear to the Romans as well as those who placed St. Peter's statue at the top of the column. It elevates the person being honored, exalting him or her above the everyday world below. Although certainly not religious in the conventional sense, such exaltation is a form of worship.

Trajan's Column is a spire built with successive marble drums placed one upon the other. Bas-relief sculpture winds around the column and spirals upward. The sculptures depict images of war and Roman conquest. It is the vertical line, however, which is fundamental to the visual impressiveness of this and other victory columns. The viewer's eye is drawn upward, with the implicit association of the subject with the heavens. Trajan's Column, and statues atop a spire in general, are also massive and ornamented. Visual persuasion here, then, is the result of the architectural elements of site, line, mass, and ornamentation, with the addition of off-white marble to lend impressiveness. One needs only to travel a short distance north of Rome to find a perhaps equally well-known victory column. This one is the Siegessäule, which now stands in roughly the middle of Berlin's Tiergarten at a confluence of streets much like those surrounding the Arc de Triomphe. Indeed, the name of this location is the Grosser Stern or Great Star.

The Siegessäule or Victory Column was built to commemorate the victories of Prussia over Denmark, Austria, and France in the so-called wars of unification. Originally, it stood in the Königsplatz in front of the Reichstag, but it was removed to its present location in the Tiergarten during the Nazi era. Of course, its site contributes strongly to the monument's visual power, and it is massive in its dimensions. The column is considerably more elaborate and ornate than Trajan's Column. It rests on a large rectangular base of red granite upon which is a circular colonnaded rotunda. The columns are smooth with a modified Corinthian design. The central column rises from the middle of the rotunda and is constructed of four sandstone drums, the base of each of which is encircled by gold garland. Atop the spire, there is a gilded statue of Victoria, the Goddess of Victory. In addition to the site and size of the Siegessäule, its red color set off by the gold garlands and the statue of Victoria itself contribute importantly to the monument's visual power.

Fig. 12-18 Siegessäule

Should one continue from the Grosser Stern east along Strasse des 17 Juni to Pariser Platz, he or she would arrive at the Brandenburger Tor with the Reichstag on the left. This famous gate marked the dividing line between West and East Germany during the Cold War following World War II. Passing through the gate, the street becomes Unter den Linden, with the Hotel Adlon on the right. Not technically a spire with a statue on top, the gate nonetheless has the kind of iconic visual power of the other structures assessed here. Indeed, it may well be the city's most familiar icon.

At the time when Berlin was a walled city, the Brandenburg Gate was the main entrance. Now, of course, there is neither a wall around the city nor one which divides it. Unter den Linden, the linden tree lined boulevard leading to the Emperor's palace, now demolished, leads away from the gate to the east. The structure is Neoclassical and modeled after the Propylaea, the entrance to the Acropolis (http://www.stadtentwicklung.berlin.de). There are five passages through the gate flanked by six fluted columns of modified Doric design. Above the columns is a traditional Greek entablature with triglyphs and metopes sculpted with bas-relief scenes. A central frieze of bas-relief figures rises above the entablature to the roof and is topped with a *quadriga* driven by the Goddess of Victory. The gate thus forms a nice counterpoint to the Siegessäule, which is on a direct line to the west. The gate was intended to symbolize peace, but it has of course come to symbolize much more, such as the divided city of the Cold War. Its visual power is certainly the result of the monument's location in the nerve center of Berlin, but it is also the result of the gate's considerable size, its ornamentation, its symmetry, and the rhythmic repetition of its colonnade.

Fig. 12-19 Brandenburg Gate

Away to the northwest and across the English Channel, the next monument to be considered here is the Nelson Memorial in London's Trafalgar Square. Vice Admiral Horatio Nelson was the hero of the Battle of Trafalgar where the British fleet defeated that of Napoléon. Nelson died in the battle, which undoubtedly contributed to the desire of the British people to memorialize his life. Many monuments to Nelson may be found in the United Kingdom, but few can match the architectural power of the one in Trafalgar Square. The site itself is among the most iconic in the city of London. Trafalgar Square is surrounded by many important structures, including the National Gallery, the National Portrait Gallery, and the Church of St. Martin in the Fields. Additionally, the square is located uphill from the Palaces of Westminster, the seat of the British government.

The column built to honor Admiral Nelson is a simple fluted Corinthian column of granite with a sandstone statue of Nelson in full battle dress standing atop it. The column is erected on a square pedestal at the four corners of which are reclining bronze lions. Trafalgar Square and its fountains radiate outward behind and to the sides of the column. Nelson's statue faces south toward Whitehall. As noted, the site is auspicious, the column is massive, and the simple elegance of the single Classical spire is visually compelling. The lions add ornament to a strong vertical line. It is probably not unintentional that one gets the impression that the late Vice Admiral continues to watch over the nation.

Fig. 12-20 Nelson Column

The last monument to be considered here is arguably not eligible for categorization as a statue atop a spire. However, it has become perhaps the strongest symbol of freedom in the world. This is the Statue of Liberty, which stands on Liberty Island in the harbor of New York City. The statue was a gift to the US by the French people. It is a massive bronze rendering of Liberty holding a tablet with the date of American independence in one hand and raising aloft a torch with the other. Her hair is crowned with a garland of seven radiating spokes symbolizing the seven seas and seven continents. The pedestal on which Liberty stands is of stone with windows framed by pilasters on its sides. It rests on a square base within a larger square base atop a star like foundation. Its site, which is not only at the entrance to America's greatest city but also near Ellis Island through which immigrants passed, is a powerful element in the statue's visual impact. Equally important is the sheer size of the structure and its green coloring.

Fig. 12-21 Statue of Liberty

Monuments and memorials are a fitting place to end this assessment of the use of architectural elements to make structures visually compelling or visually persuasive. They represent a synthesis of architecture and art whose intent to commemorate is unquestionably a persuasive one. The motivations of remembrance and celebration are as old as humankind. On a smaller scale than the monuments discussed here are the countless cemetery headstones and mausoleums throughout the world. Ancient oratory was divided into legislative or deliberative speeches, forensic or legal speeches, and epideictic speeches, which were of praise or blame. Although epideictic speeches are today not so prominent in the orator's repertoire, there is no lack of interest in the construction of monuments. Indeed, much is said by various groups about the lack of monuments to honor some leader or event, and the pace of construction of such memorials in Washington, DC, for example, has quickened in recent years rather than abated.

AFTERWORD

SOME REFLECTIONS ON THE SOCIAL INFLUENCE OF ARCHITECTURE

This book has utilized basic elements of architectural design to assess the visual persuasiveness of many kinds of structures principally in the Western world: rhythm, line, scale, mass, site, light, color, ornament, acoustics, and the like. Underlying this use of elements has been the idea that it is its visual elements which account for architecture's compelling form and that, like the arguments and other characteristics of verbal persuasion, these elements make up a structure's visual power. In Chapter Two, these elements were considered in terms of how they have been combined by architects into construction forms, such as arches, columns, domes, entablatures, pediments, and so on, that make up the numerous styles that have evolved throughout history. At its heart, then, this book has been about the daily and ubiquitous influence of buildings, on people's views of their homes, their cities, their countries, and surely also as a result their senses of individual well being. This is the sense in which one may legitimately say that architecture has social influence, that it is visually persuasive.

This type of social influence is not, of course, identical to that of verbal persuasion. The latter is, as noted earlier, discursive. It is a means of advocacy, and propositions are its stock in trade. "You should stop smoking," a persuasive speaker might say, or "you should invest your money with this investment group." Visual persuasion, by contrast, does not offer a proposition for consideration but instead affects the receiver with such powerful stimuli—often such incomparable beauty—that he or she is stunned, or overwhelmed, or delighted, or compelled. All of these and other terms used to express the outcome of seeing a compelling form refer, of course, to emotional responses. This does not mean necessarily that there are no intellectual components to such responses, only that they are nondiscursive and rather difficult to quantify and sometimes even to define. While they are not entirely abstract, they are certainly complex. Because of this, such terms are troublesome to social scientists, and they may often be avoided for the same reason. However, it should be abundantly clear to the reader that there are not concrete, objective terms readily available to describe the range of responses one may have to visual stimuli. Responses to verbal persuasive attempts may quite reliably be measured. Not so with visual ones, but to say that there has been no effect would be terribly misleading.

Visual stimuli may accompany verbal persuasion, as in the case of the environment surrounding a speech or a speaker's visual aids. Similarly, verbal stimuli may accompany visual persuasion, as in the case of a descriptive caption or slogan. The two forms of persuasion most often do complement each other, but receivers clearly respond differently to propositional advocacy and to compelling form. The two forms use different means. Verbal persuasion is built on arguments which appeal to reason, emotion, and to the speaker's credibility. Visual persuasion is built on the basic elements of visual literacy—lines, dots, geometric forms, and the like—which appeal to one's experience of beauty, although that term too is fraught with definitional trouble. However, visual persuasion affects people in groups, hence it is social. It enhances their well-being, their civic and personal pride, and their comfort, hence it is a means of influence.

This volume is the fourth in a series that began with a general examination of structures as argument. In that sense, the present volume is an elaboration on the first. The second and third volumes concentrated primarily on art museums, which are surely the richest vein to mine for evidence of visual persuasion. Art museums are impressive as structures, but they also contain collections that have been assembled to have a particular effect on viewers. Finally, these collections are carefully arranged and displayed to have one effect on the viewer and not another.t In a sense, this volume exhausts the structures that might be assessed for evidence of humankind's aspirations to beauty and thus to visual influence. To be sure, there are undoubtedly other points of view to be offered on the same material. However, there are other efforts by mankind to fashion the world in a pleasing and influential visual way, and the first such efforts that come to mind are the exterior spaces that have been designed for visual effect. Hopefully, another volume will address the gardens, parks, zoos, preserved battlefields, and the like that illustrate this purpose.

REFERENCES

A Forest of Pillars. http://www.nytimes.com/2005/05/09/arts.

Anderson, W. 1985. *The rise of the Gothic.* Salem, NH: Salem House.

Andrews, M. 2006. *Historic Texas courthouses.* Albany, TX: Bright Sky Press.

Answers.com. Home. http://www.answers.com.

Architecture in Fort Worth. Home. http://www.fortwortharchitecture.com.

Archiseek. Home. http://archiseek.com.

A View on Cities. Home. http://www.aviewoncities.com

Babbitt, S. 1995. *Philadelphia Museum of Art: Handbook of the collections.* Philadelphia: Philadelphia Museum of Art.

Berlo, D. K. 1960. *The process of communication.* New York: Holt, Rinehart and Winston.

Buchler, J., ed. 1955. *Philosophical writings of Peirce.* New York: Dover Publications.

Burke, K. 1945. *A grammar of motives.* New York: Prentice-Hall.

Blumenson, J. J.-G. 1977. *Identifying American architecture: A pictorial guide to styles and terms, 1600-1945.* Nashville, TN: American Association for State and Local History.

Buerger, D., and Z. Omar, eds. 2001. *Eyewitness travel: New Zealand.* London: Dorling Kindersley.

Building SHSU. Home. http://buildingshsu.com.

Château de Vincennes. Home. http://en.chateau-vincennes.fr.

Ching, F. D. K. 1979. *Architecture: Form,space, & order.* New York: Van Nostrand Reinhold Company.

Cole, E., ed. 2005. *The grammar of architecture.* New York: Metro Books.

Costecalde, C.-B. 1997. *A short tour of St. Patrick's Cathedral.* Strasbourg, FR: Éditions du Signe.

Cowan, H. J. 2002. Transport and Communications. In *A guide to the world's greatest buildings: Masterpieces of architecture and engineering,* ed. T. Howells, 182-215. San Francisco: Fog City Press.

Dallas Architecture Info. Home. http://www.dallasarchitecture.info.

Designbuild-network.com. Home. http://designbuild-network.com.

Dondis, D. A. 1973. *A primer of visual literacy.* Cambridge, MA: MIT Press.

Duncan, C. 1995. *Civilizing rituals: Inside public art museums.* London: Routledge.

E-Architect. Home. http://www.e-architect.co.uk.

Ellmerer, B. 1993. *Schloss Schönbrunn.* Vienna: Schloss Schönbrunn.

Field, D. M. 2007. *The world's greatest architecture past and present.* Edison, NJ: Chartwell Books.

Fleming, J., H. Honour, and N. Pevsner. 1991. *The Penguin dictionary of architecture,* 4th Ed. London: Penguin Books.

Ford, E. H. 1998. *Washington, D.C. from $60 a day.* New York: Simon & Schuster.

French Creole Architecture. Home. http://www.nps.gov/history/nr/travel/Louisiana/architecture.htm.

Gayford, M., J. Holden, R. Moore, C. Smith, J. Snow, and T. Travers. 2005. *Tate Modern: The first five years.* London: Tate Trustees.

Glancey, J. 2006. *Architecture.* London: DK Publishing.

Global Architecture Encyclopedia. Home. http://www.glasssteelandstone.com.

Go There Guide. Home. http://www.gothereguide.com.

Guide to Castles of Europe. Home. http://www.guide-to-castles-of-Europe.com.

Hanna, B. 2002. Monuments and Memorials. In *A guide to the world's greatest buildings: Masterpieces of architecture and engineering,* ed. T. Howells, 218-245. San Francisco, CA: Fog City Press.

Harvard University. Home. http://www.harvard.edu.

Hein, H. 2006. *Public art: Thinking museums differently.* Lanham, MD: Alta Mira Press.

Historic Royal Palaces. Home. http://www.hrp.org.uk.

History of Gothic Architecture. http://web.kyoto-inet.or.jp/org.

Hotel Adolphus. Home. http://www.hoteladolphus.com.

Howells, T. 2002. *A guide to the world's greatest buildings: Masterpieces of architecture and engineering.* San Francisco: Fog City Press.

Humboldt University. Home. http://www.hu.berlin.de

Huxtable, A. L. 1984. *The tall building artistically reconsidered: The search for a skyscraper style.* New York: Pantheon Books.

Italy Guides.it. Home. http://www.italyguides.it.

Lee-Fendall House. Home. http://www.leefendallhouse.org.

Luminarium: Encyclopedia Project. Home. http://www.luminarium.org/encyclopedia.

Macdonald, S. 1996. Introduction. In *Theorizing museums: Representing identity and diversity in a changing world,* ed. S. Macdonald and G.

Fyfe, 1-18. Oxford, UK: Blackwell Publishing/The Sociological Review.

Magi, G. 1990. *Tout Paris.* Firenze: Bonechi.

McAlester, V. & L. 1985. *A field guide to American houses.* New York: Alfred A. Knopf.

Merton College. Home. http://www.merton.ox.ac.uk.

Messaris, P. 1997. *Visual persuasion: The role of images in advertising.* Thousand Oaks, CA: Sage.

Moe, A. 2006. Foreword. In *Historic Texas courthouses,* M. Andrews, 16-19. Albany, TX: Bright Sky Press.

Museum of Contemporary Art, Sydney. Home. http://www.mca.com.au.

National Cathedral. Home. http://www.nationalcathedral.org.

Palladio, A. 1965. *The four books of architecture.* Trans. I. Ware. New York: Dover Publications.

Paris Travel Guide. Home. http://www.paris.com.

Petty, R. E., and J. T. Cacioppo. 1986. The Elaboration Likelihood Model of persuasion. In *Advances in experimental social psychology,* ed. L. Berkowitz, 19:123-205. New York: Academic Press.

Philadelphia Museum of Art. Home. http://www.philamuseum.org.

Provence & Beyond. Home. http://www.provencebeyond.com.

Quintin Lake Photography Blog. Home. http://blog.quintinlake.com/2010/4/26.

Ragsdale, J. D., ed. 2007. *Structures as argument: The visual persuasiveness of museums and places of worship.* Newcastle, UK: Cambridge Scholars Publishing.

Ragsdale, J. D. 2009a. *American museums and the persuasive impulse: Architectural form and space as social influence.* Newcastle, UK: Cambridge Scholars Publishing.

—. 2009b. *Western European museums and visual persuasion: Art, edifice, and social influence.* Newcastle, UK: Cambridge Scholars Publishing.

Rees, J. C. 2001. Making the most of your visit. In *Mount Vernon official guidebook.* Mount Vernon, VA: The Mount Vernon Ladies Association of the Union.

Rhodes, G., F. Proffitt, J. M. Grady, and A. Sumich. 1998. Facial symmetry and the perception of beauty. *Psychonomic Bulletin & Review, 5,* 659-669.

Roth, L. M. 1993. *Understanding architecture: Its elements, history, and meaning.* New York: Harper Collins.

Royal Albert Hall. Home. http://www.royalalberthall.com.

Ruskin, J. 1989. *The seven lamps of architecture.* New York: Dover.

Sacred Destinations. Home. http://www.sacred-destinations.com.
Salisbury Cathedral. Home. http://www.salisburycathedral.co.uk.
Sam Houston State University. Home. http://www.shsu.edu.
Samoyault, J.-P. 1994. *Fontainebleau: A guide to the visit.* Montreuil: Mahé.
Schleier, M. 1990. *The skyscraper in American art: 1890-1931.* New York: Da Capo Press.
Scott, R. A. 2005. *The gothic enterprise; A guide to understanding the medieval cathedral.* Berkeley: University of California Press.
Senate Department for Urban Development. Home. http://www.stadtentwicklung.berlin.de.
Stein, S. R. 1997. Jefferson's Monticello. In *Monticello: A guidebook.* Charlottesville, VA: Thomas Jefferson Foundation.
Strickland, C. 2001. *The annotated arch: A crash course in the history of architecture.* Kansas City, MO: Andrews McMeel Publishing.
Sydney Opera House. Home. http://www.sydneyoperahouse.com.
Symmes, P. 2010. History in the making. *Newsweek,* 1 March.
Tabraham, C. 2008. *Edinburgh castle: The official souvenir guide* Edinburgh: Historic Scotland.
Tate. Home. http://www.tate.org.uk.
Te Papa Tongarewa. Home. http://www.tepapa.gov.nz.
Texas State Preservation Board. Home. http://www.tspb.state.tx.us.
The White House. Home. http://www.whitehouse.gov.
Time Out. Home. http://www.timeout.com.
Tresidder, J. 2006. *Symbols and their meanings.* New York: Barnes and Noble.
Van Leeuwen, T. A. P. 1988. *The skyward trend of thought: The metaphysics of the American skyscraper.* Cambridge, MA: The MIT Press.
Venclova, T. 2002. *Vilnius city guide.* 3rd Ed. Vilnius, LT: R. Paknio.
Vitruvius, 1960. *The ten books on architecture.* Trans. M. H. Morgan. New York: Dover Publications.
York Minster. Home. http://www.yorkminster.org.

INDEX

Acoustics, 7, 67
Admiral Nelson, 286
Adolphus Hotel, 237, 239, 240
Aigues-Mortes, 198, 199, 200
Ali'iōlani Hale, 167, 182
Amon Carter Museum, 45
Arc de Triomphe, 12, 26, 47, 254, 273, 278, 279, 284
architecture, 31, 39, 47, 49, 291, 293
art, 3, 39, 41, 49, 50, 291
Art Gallery of New South Wales, 51, 55
Ashmolean Museum, 141, 142, 146, 147
Austin Hall, 161, 162
Austrian Parliament, 172, 173
Bank of America Plaza, 250
Banqueting House, 33, 71, 75, 76
baroque, 33, 34, 35
Basilica di Santa Maria del Fiore, 31, 32, 96, 98
Basilica of San Marco, 28, 100, 211
Bass Concert Hall, 82
Bath Abbey, 119, 120
beauty, 45
Belgian Royal Palace, 207, 208
Bell County, 184, 185
Bodleian Library, 21, 72, 73, 141, 144, 145, 146
Brandenburg Gate, 35, 171, 279, 285
Bridge of Sighs, 211, 256
British Museum, 2, 15, 24, 35, 39, 40, 46, 47, 48, 50, 146
Brooklyn Bridge, 244, 259, 260
Buckingham Palace, 215, 216, 217, 269
Byzantine architecture, 27, 100, 101
Cambridge University, 147, 148, 156
Carcassonne, 199, 200, 201
Carlyle House, 227, 228
Carpenter Center for the Visual Arts, 158, 159
Castel sant'Angelo, 200, 203
Cathedral of Santa Maria la Real de la Almudena, 103
Cathedral of St. Barbara, 105
Cathedral of St. Michael and St. Gudula, 95
Cathedral of St. Peter, 101, 107, 108, 109
Cathedral of St. Stephen, 110
Cathedral of St. Vitus, 104
Centre Georges Pompidou, 39, 49
Chain Bridge, 173, 213, 257, 259, 260
Charles Bridge, 243, 257, 258, 259
Château de Versailles, 34, 204
Château de Vincennes, 198, 290
Chiang Kai-Shek Memorial, 266, 272
Christchurch Art Gallery, 51, 60
ChristChurch Cathedral, 136
Chrysler Building, 244, 246, 247
Circus, 101, 119, 234, 235
Classical architectural style, 21, 22
Colosseum, 11, 26, 62, 63, 64, 72, 283
Comerica Bank Tower, 250
Congress of Deputies, 167, 174, 175
Convent of San Marco, 32
Corinthian, 22, 23, 24, 27, 31, 42, 73, 76, 78, 94, 97, 106, 112, 116, 117, 121, 127, 149, 153, 154, 155, 157, 172, 173, 175, 177, 181, 184, 185, 188, 208,

210, 212, 213, 216, 217, 225, 234, 272, 284, 286
D. R. Horton Tower, 252
Dallas County, 187, 188, 189
Denton County, 187, 190
discursive, 89
Doge's Palace, 31, 100, 211, 212, 256
Doric, 10, 21, 22, 23, 24, 74, 104, 107, 119, 120, 153, 157, 161, 179, 181, 188, 192, 205, 207, 215, 221, 223, 230, 234, 236, 266, 285
Durham Castle, 122, 151, 196, 197
Durham Cathedral, 122, 123, 151, 196
Edinburgh Castle, 123, 152, 169, 197, 215
Eiffel Tower, 79, 265, 273, 277, 278
Elaboration Likelihood Model, 3, 292
Ellis County, 186, 187
Empire State Building, 15, 244, 245, 246, 247
Esztergom Cathedral, 105, 106
FDR Memorial, 265, 281, 282
Flatiron Building, 244, 245
Flinders Street Station, 133, 262, 263
Fontainebleau, 33, 205, 206, 207, 208, 293
Fountain Place, 249
Frauenkirche, 109, 110, 111
Frick Mansion, 226
Gaertner Performing Arts Center, 66, 67, 69, 70
Galleria degli Uffizi, 31
Gare d'Orsay, 263
Giotto's campanile, 99
Globe Theatre, 49, 71, 72
Göbekli Tepe, 86
Golden Gate Bridge, 243, 244, 260, 261
Gothic style, 29, 30, 88, 90, 91, 92, 94, 109, 110, 115, 122, 123, 125, 131, 142, 144, 161, 169, 179, 191, 268, 270
Grace Cathedral, 124, 128, 129
Grand Hotel, 237, 238, 239, 240
Harbor Bridge, 65, 66, 261
Harvard University, 149, 155, 156, 157, 291
Hill County, 185, 186, 190
Hohensalzburg, 200, 201, 202
Hohenschwangau, 208, 210, 211
Houmas House, 229, 230, 231
Humboldt University, 155, 291
Hungarian Parliament, 173, 174, 259
Hungarian Royal Palace, 213
Ian Potter Centre, 54, 55
icon, 3, 4, 38, 49
Imperial College of Science and Technology, 150
index, 3, 4
Iolani Palace, 182, 220, 224, 225
Ionic, 10, 20, 22, 23, 24, 33, 42, 43, 48, 55, 75, 76, 104, 109, 145, 146, 151, 153, 157, 161, 182, 209, 222, 224, 227, 234, 235, 267
Italian Renaissance architecture, 31
Jefferson Memorial, 267, 268, 281
JPMorgan Chase Tower, 250, 251
Kennedy Center for the Performing Arts, 80
Kimbell Art Museum, 41, 45
King William historic district, 233
King's College Chapel, 148, 169
Korean War Veterans Memorial, 265, 266, 280, 281
La Grande Arche de La Défense, 254
Le Mont St. Michel, 92, 93, 94, 95
Leaning Tower of Pisa, 97, 98
Lee's boyhood home, 228
Lee-Fendall House, 228, 229, 291
Lincoln Center for the Performing Arts, 80
Lincoln Memorial, 176, 265, 266, 267, 274, 275, 281

Louvre, 2, 12, 24, 25, 34, 39, 40, 42, 46, 47, 48, 50, 83, 160, 198, 204, 254, 263
Massachusetts Hall, 157, 158
Matthias Church, 106, 107, 110
Melbourne Museum, 51, 52, 53, 54, 60
Memorial to the Murdered Jews of Europe, 273, 278, 280
Metropolitan Museum of Art, 39, 41
Meyerson Symphony Center, 83
Modern Art Museum of Fort Worth, 45
Monticello, 33, 220, 222, 223, 293
Mormon Tabernacle, 124, 127
Mormon Temple, 128
Mount Vernon, 33, 220, 221, 222, 223, 292
Museum of Contemporary Art., 55, 56
National Cathedral, 35, 124, 125, 126, 127, 292
National Gallery of Victoria, 51, 53, 54, 61, 79
National Monument to Vittorio Emanuele II, 271
Neoclassical, 35, 42, 43, 55, 72, 103, 105, 106, 107, 108, 117, 142, 144, 145, 146, 149, 151, 153, 157, 161, 171, 172, 173, 175, 176, 177, 179, 181, 192, 207, 212, 214, 217, 223, 285
Neuschwanstein, 208, 210
Notre-Dame de Chartres, 91, 92, 107, 115
Notre-Dame de Paris, 3, 87, 90, 91, 92, 95, 115, 127
Notre-Dame de Reims, 92, 94
Oak Alley, 229, 230, 231
Oakwood subdivision, 232
Old St. Paul's, 134, 135
Opéra Garnier, 76, 77, 78, 80
Oxford University, 72, 141, 142, 146
Painted Ladies, 233, 234
Palace of Holyroodhouse, 123, 169, 215
Palace of Westminster, 35, 76, 168, 173, 175
Palacio Real, 103, 104, 212, 213
Palazzo Pitti, 31
Parliament House, 167, 178, 179, 180, 181
Parliamentary Library, 179
persuasion, 39, 48, 292
Philadelphia Museum of Art, 24, 39, 41, 43, 44, 290, 292
Place des Vosges, 234, 236, 237, 238
Pont du Gard, 26, 244, 255, 256, 257
Prague Castle, 104, 200, 202, 258
presentation, 3
Prince Albert Memorial, 265, 268
Radcliffe Camera, 144, 145, 146
Reichstag, 167, 171, 172, 279, 284
Renaissance architecture, 31, 32, 33, 76, 97, 257
Reunion Tower, 251
Rialto Bridge, 244, 256, 257
Rijksmuseum, 39
RMIT University, 163
rococo, 34, 35
Romanesque style, 29, 30
rose window, 89
Royal Albert Hall, 71, 73, 74, 84, 150, 236, 268, 292
Royal Crescent, 119, 234, 235, 236
Sainte-Chapelle, 92, 198
Salisbury Cathedral, 11, 118, 293
Sam Houston State University, 66, 72, 138, 159, 161, 162, 293
Schönbrunn, 11, 208, 209, 216, 217, 291
Scottish Parliament, 169, 170, 191, 215
Senate House, 148, 149
Shadows on the Teche, 229, 231
Sheldonian Theatre, 72, 73, 141, 142, 144

Siegessäule, 171, 265, 266, 283, 284, 285
Sir Walter Scott Memorial, 265, 270
Sorbonne, 141, 154
South Kensington, 150, 151, 234, 236, 237
St. Andrew's Cathedral, 131
St. Denis, 90, 91
St. Giles, 123, 124, 152, 169
St. Martin in the Fields, 117, 157, 285
St. Mary's Cathedral, 130
St. Pancras Station, 262
St. Patrick's Cathedral, 124, 126, 127, 131, 132, 290
St. Paul's Cathedral, 34, 49, 113, 115, 116, 132, 133, 135, 176
State Capitol Building in Austin, 183
State Library of Victoria, 167, 180, 181
Statue of Liberty, 286, 287
Stockholm Royal Palace, 214, 215
Suppression of the Monasteries, 113
Sydney Opera House, 56, 64, 65, 66, 79, 262, 293
symbol,, 4
Tarrant County, 187, 188, 189
Tate Modern, 39, 46, 49, 50, 56, 72, 291
Te Papa Tongarewa, 51, 58, 178, 293
Texas A&M University, 159, 232
The Pickle, 193, 253
Tower Bridge, 193, 243, 257, 258
Town Hall, 33, 167, 191, 192, 193
Trajan's Column, 25, 283, 284
Transamerica Building, 248
University of Durham, 151
University of Edinburgh, 152, 153, 169
US Capitol, 17, 116, 167, 176, 183, 273
US National World War II Memorial, 275
US Supreme Court, 167, 177
USS Arizona, 276
Victoria Memorial, 269
Victoria Station, 216, 262
Victoria University, 164
Victorian Arts Centre, 53, 78, 79
Vietnam Veterans, 265, 266, 273, 274, 281
Vilnius Cathedral, 107, 108, 109
Visual literacy, 6, 8
Vitruvius, 6, 7, 9, 10, 14, 15, 16, 22, 23, 62, 76, 293
Washington Monument, 176, 265, 273, 274, 275
Wellington Cathedral, 134, 135
Westminster Abbey, 113, 114, 116, 168
White House, 125, 220, 221, 223, 224, 240, 293
White Tower, 197
Willis Tower, 247, 248
Winspear Opera House, 83, 84
York Minster, 121, 122, 126, 131, 293